System Dynamics for Performance Management

Volume 1

Series Editor

Carmine Bianchi, CED4-System Dynamics Group, University of Palermo, Palermo, Italy

More information about this series at http://www.springer.com/series/13452

Carmine Bianchi

Dynamic Performance Management

CED⁴
System Dynamics Group

Springer

Carmine Bianchi
CED4-System Dynamics Group
University of Palermo
Palermo
Italy

ISSN 2367-0940   ISSN 2367-0959   (electronic)
System Dynamics for Performance Management
ISBN 978-3-319-31844-8   ISBN 978-3-319-31845-5   (eBook)
DOI 10.1007/978-3-319-31845-5

Library of Congress Control Number: 2016934960

Printed on acid-free paper

This Springer imprint is published by Springer Nature
The registered company is Springer International Publishing AG Switzerland

To Lilla

Preface

This book is the first in a series on "System Dynamics for Performance Management." The purpose of the series is to bridge the gap between system dynamics and its applications in organizations, with a specific focus on performance management.

Most contributions to this series are expected to come from the research and teaching at the doctorate level program in "Model Based Public Planning, Policy Design and Management," which was started in 2007 as a double degree collaboration between the Universities of Palermo and Bergen. The collaboration is further enhanced by the joint "European Master in System Dynamics," which is delivered as an "Erasmus Mundus" EU-funded program with the Radboud University of Nijmegen and the University of Lisbon. The teaching work related to the Master in Public Management, recently started by the System Dynamics Group at the University of Palermo, also will contribute to such endeavor.

The purpose of the book is to support the reader's understanding of how to design and implement planning and control (P&C) systems that can help organizations manage their growth and restructuring processes in a sustainability perspective. More specifically, the book shows how to develop system dynamics models that can better support an understanding of the following:

- What is organizational performance and how to frame and measure it;
- How to identify and map the processes underlying performance;
- How to design and implement a dynamic performance management system and link it to strategic planning;
- How to tie strategic resource dynamics to processes and performance indicators;
- How to link strategic resources and performance indicators to responsibility and incentive systems.

The book builds on my work over the last 20 years in research, teaching, and consulting in the field of system dynamics modeling applied to performance management, with a specific focus on public sector organizations and business.

It also includes numerous case studies and dynamic performance management models for providing examples of how such methods work in practice.

Reading the book is beneficial to: (1) graduate students in business and public management, (2) researchers in performance management and system dynamics modeling, and (3) managers in business and public sector organizations in the fields of performance management and planning.

The book is not designed to enable the reader to become an experienced system dynamics modeler; rather, it aims to develop the reader's capabilities to design and implement performance management systems by using a system dynamics approach. A literature review is included to provide a guideline for further improvements to those readers who wish to develop relevant, specific, and detailed system dynamics modeling skills and to establish the foundation for teaching system dynamics applied to performance management in organizational and inter-organizational contexts. This is particularly relevant for those graduate students who have taken system dynamics courses and need to apply their own skills to business and public management.

To this end, system dynamics students are often involved in case studies. In such endeavors, most of them are often inclined to "fit" the generic stock-and-flow feedback structures and the model examples they have previously learned in system dynamics classes to the issues framed by the cases. Based on such skills, they may immediately look for the dynamic problem to model and the data from which key-variables reference behavior modes could be identified. Though such structured approach can be useful to help students in implementing the methodological and technical skills they have previously acquired, there may be a risk to bound the "potential market" for such skills to only those organizations which have specific staff specialized in system dynamics.

The field of performance management provides a potentially much wider "market" for system dynamics skills, also in those organizations that do not have a "system dynamics" staff and culture. In such contexts, the design and implementation of a "dynamic" performance management system, and the identification (at a corporate and departmental level) of performance drivers may foster dynamic problem finding. A system dynamics model that "speaks" the same language as that of people in performance management is the next step of this process.

Such approaches also are beneficial to performance management analysts, enabling them frame their professional field within the broader context of the system. Using a dynamic performance management approach can improve an organization's capability to: (1) understand and manage the forces driving performance over time, (2) set goals and objectives that may properly and selectively gauge results and match them to the key responsibility areas in the planning process, (3) perceive the effect of physical and perception delays on organizational performance, (4) frame and measure the indirect effects of intangibles on performance, (5) detect and counteract possible unintended effects of performance measurement on group and individual behavior, and (6) foster a learning-oriented—rather than bureaucratic/incremental approach—to planning.

This overcomes the risk that P&C systems are designed and used according to a mechanistic and static perspective. Such risk may generate an *illusion of control*, rather than an enhanced capability of organizational decision-makers to manage sustainable development, to promptly detect symptoms of crisis, to look for the causes of financial results, to set sustainable restructuring policies, and to search for consistency in different subsystems, sectors, departments, or governmental functions.

The book consists of five chapters. The first chapter introduces the topic of dynamic performance management and the need for such approach. Basic system dynamics concepts are introduced and linked to the performance management field. Chapters 2 and 3 respectively frame the field of dynamic performance management and illustrate how to operationalize dynamic performance management. Chapters 4 and 5 respectively illustrate how to use dynamic performance management in the public sector and business.

I trust that this work will become a bridge between two fields of knowledge and practice (i.e., system dynamics and performance management) from whose interplay organizations can be better supported in their efforts toward sustainable growth and lifelong endurance. I would like to thank many friends and colleagues for the support they—either directly or indirectly—gave me in the process leading to this book.

I am indebted to Vittorio Coda and Carlo Sorci, respectively from Bocconi and Palermo University, whose ideas have strongly influenced, since the mid-80s, the way I could frame the linkages between performance, organizational development, and sustainability. I also am indebted to Elio Borgonovi, from Bocconi University: his scientific production has significantly affected my conceptual framework on performance and sustainable development in public sector organizations.

I wish to thank Pål Davidsen, from the University of Bergen. Since 1995, I have shared with Pål innovative teaching projects, which enabled me to set up and develop a System Dynamics Group in my department. I'm indebted to John Morecroft and Kim Warren, respectively from London Business School and Strategy Dynamics LTD: their research affected my view on strategic resource dynamics and their effects on organizational performance. Also the research I carried out with Graham Winch, from Plymouth University Business School, has greatly contributed in this regard.

I also wish to thank the graduate students and the researchers of CED[4] System Dynamics Group at my department, Enzo Bivona, Federico Cosenz, and our Ph.D. students, with whom I could share ideas, teaching and research experiences that contributed to this book. I wish to pay special thanks to my good friends, Bill Rivenbark and Salvatore Rotella, respectively from the University of North Carolina at Chapel Hill and Riverside College. They both patiently supported me in reviewing and criticizing the book. Also, the research ideas and teaching experiences in the field of public performance management I have carried out with Bill greatly contributed to both improve the conceptual framework and the practical relevance of this book.

Last but not least, I wish to thank Lilla, Floriana, and Alessandro. Lilla's patience and support made writing this book possible. This book is for her.

Contents

About the Author

Carmine Bianchi is Professor of Business & Public Management at the University of Palermo (Italy), where he is the scientific coordinator of CED^4 System Dynamics Group (www.ced4.com). He is the Director of the doctoral level program in "Model Based Public Planning, Policy Design, & Management," and of the master level program in Public Management.

Professor Bianchi is member of the Steering Committee of the "European Master in System Dynamics," a joint degree among the Universities of Palermo, Nijmegen (the Netherlands), Bergen (Norway), and Lisbon (Portugal). This is an "Erasmus Mundus" funded program, sponsored by the European Commission.

He has published in numerous academic and professional journals. He also serves on the Scientific Committee of various academic publications.

Professor Bianchi has an extensive international research and consulting experience with public and private sector organizations. Such activities consist in the design of policies and the outline of programs linking strategy and implementation. Consulting and education projects that Prof. Bianchi has undertaken include: strategic planning and control, performance management and reporting, as well as system dynamics modeling for performance improvement and crisis prevention (Dynamic Performance Management).

In the last decade, Prof. Bianchi has been strengthening an international network related to "Dynamic Performance Management." He has been collaborating with many universities throughout the world, from Europe (Bergen, Norway; Nijmegen, The Netherlands; St. Gallen, Switzerland; KTH, Stockholm, Sweden), to Australasia (Top Education Institute, Sydney; Multimedia University and National University of Malaysia, Kuala Lumpur; NAPA—Hanoi; Rangsit University, Bangkok; Universitas Indonesia, Jakarta; National Academy of Public Administration—NAPA, Hanoi), and America (University of Campinas—UNICAMP, Brazil).

In the USA, Prof. Bianchi has collaborated extensively with the School of Government of University of North Carolina at Chapel Hill. More recently he has undertaken projects with the Center for Human Resource Management State, Stony Brook University, with the School of Public Affairs at the Baruch College, New York, and with the College of Public Affairs at the University of Baltimore.

He has been also visiting professor at the Rockefeller College of Public Affairs & Policy, University at Albany (USA), at Top Education Institute, Sydney (Australia), and at Unicamp, Faculty of Technology, Limeira (Brazil).

He has also run professional workshops in: Africa (Tripoli, Tunis), Australasia (Bangkok, Brisbane, Hanoi, Jakarta, Kuala Lumpur, Penang, Singapore), Europe (Istanbul, London, Milan, San Gallen, Stockholm), and Latin America (Buenos Aires—Argentina, Brasilia, Sao Paolo—Brazil, Colima—Mexico).

Professor Bianchi can be contacted at: bianchi.carmine@gmail.com or carmine.bianchi@unipa.it.

Chapter 1
Managing Organizational Growth and Dynamic Complexity

1.1 Complexity, Decision-Making and Human Error

Managers, entrepreneurs and administrators are often reluctant to invest time in order to frame the future of their own organization. They may look too much focused on reducing the complexity of current management. The vision of the future they create in their own minds may remain unchanged even for many years, or may incrementally and ambiguously change, with no effort on learning, on perceiving the early symptoms of crises, and assessing the sustainability of organizational growth rate.

A bounded capability to frame organizational future is not only related to those contexts where goals are made explicit through structured and long-term strategic plans (Ansoff 1965). It can also be found in those organizations whose visions are imbedded in the flow of current and inertial decisions, which often emerge through a tortuous process characterized by bounded rationality, heuristics, rules of thumb (Simon 1945), political bargaining (Lindblom 1959), and irrational decision-making (Cyert and March 1963; Cohen et al. 1972). It may also emerge in those organizations adopting a decision-making process based on *incremental changes*, or on a mix of deliberate and emergent strategies.[1]

Complexity and delays between causes and effects provide major obstacles to learning, early symptom perception, and sustainable growth management, regardless the strategic process model used to frame decision-making in organizational settings.

Human beings are subject to decision-making that is often based on wrong assumptions, due to bias towards the problems at hand. They also adopt a simplistic

---

[1]In the first context, people making decisions want to "test assumptions and have an opportunity to learn from and adapt to the others' responses" (Quinn 1978, p. 9). In the second context, the strategy process is a result of both a planned and an unintentional set of actions (Mintzberg and Waters 1985).

© Springer International Publishing Switzerland 2016
C. Bianchi, *Dynamic Performance Management*, System Dynamics
for Performance Management 1, DOI 10.1007/978-3-319-31845-5_1

view of uncertainty. In fact, the less likely and infrequent events are often unperceived, and unpleasant outcomes tend to be ignored.

Dorner (1997) pointed out a number of factors related to dynamic complexity, that mostly give rise to human error in decision making, i.e.[2]:

- Linear (i.e. sequential) thinking, which disregards propagation and ramification effects between different variables affected by decisions.[3] This often leads to a misperception of feedbacks and delays, as well as to a poor recognition of non-linear growth and decay processes;
- Weak information collection and hypothesis formation[4];
- Thematic vagabonding, leading to a loss of coherence in thinking, where multiple interacting themes are treated superficially and independently, so that people may "jump" from one theme to the next without any consistency.

As Simon (1957, p. 198) remarked, "the capacity of the human mind for formulating and solving complex problems is very small compared with the size of the problem whose solution is required for objectively rational behavior in the real world of even for a reasonable approximation of such objective rationality". Bounded rationality is intrinsic in decision-making, because of human cognitive limitations. When people make decisions they portray bounded capabilities to: (1) generate a set of hypotheses including *all* alternative courses of action; (2) collect and process *all* the information that would enable them to predict the consequences that choosing a given alternative would imply; (3) make an accurate evaluation of such consequences and an objective selection among them (Morecroft 1985, p. 901).

Therefore, organizations are structured "to transform intractable problems into tractable ones" (Simon 1979, p. 501). So, organizational decision-making is rationally carried out under simplified conditions of choice. "One function that organization performs is to place the organization members in a psychological

---

[2]A similar viewpoint has been developed by Hamel and Prahalad (1994, pp. 120–121), who remark: "We often come across companies that have set an ambitious long-term goal, perhaps to double revenue and profits over five years, or to dramatically increase the proportion of revenues coming from new businesses, but have devoted almost no intellectual effort to thinking through the medium-term capability-building program that is needed to support that goal. In too many companies there is a grand, and overly vague, long-term goal on one hand ... and detailed short-term budgets and annual plans on the other hand ... with nothing in between to link the two together. ... There seems to be, in many companies, an implicit assumption that the short term and long term abut each other, rather than being dovetailed together. But the long term doesn't start at year five of the current strategic plan. It starts right now!".

[3]"When solving such complex tasks, most people are not interested in finding out the existent trends and developmental tendencies at first, but are interested instead in the 'status quo'". Dorner (1980).

[4]"Most mistakes in information collection are 'channeling errors', which result from a preformed image of reality. Subjects are not prepared to look at the whole range of information, but only to the narrow part they—according to their image of the system—consider to be important". Dorner and Schaub (1994).

environment that will adapt their decisions to the organization objectives, and will provide them with the information needed to make these decisions correctly" (Simon 1945, p. 92).

However, though this simplification may allow seemingly rational choices made by many individuals, it does not automatically guarantee that such "choices are consistent and mutually supportive" (Morecroft 1985, p. 901). For instance, factoring problems into sub-problems may simplify decision making over complex issues; however, its shortcoming is that each specialized organizational decision-maker involved in a fraction of a complex problem receives only a limited part of the available information flow. If, on the one hand, this allows a prompt evaluation and decision making in each business sector, on the other hand, it may hinder a systemic perception of the problem itself.

1.2 Framing Trade-Offs and Policy Resistance in Dynamic Complex Systems

It has been demonstrated that misperceiving dynamic[5] complexity is a main cause of poor organizational performance and crisis.[6]

Dynamic complexity should be distinguished from combinatorial (or static) complexity.[7] The latter is associated with a high number of variables and multiple relationships, where both relationships and variables tend to remain stable over time. The former is, instead, associated with unpredictability in such interconnections, because of delays, nonlinearities and multiple feedback loops whose dominance affects the system's behavior (Senge 1990; Morecroft 1998; Sterman 1994).[8]

---

[5]The term "dynamic" means changing in time and interactive with the process in which it occurs (i.e., having a feedback characteristic). Its opposite, "static," means conceptualized as abstract from the process; thus, its value can change in time, but that leads to no consequences for the process in which the static factor is found.

[6]Arie De Geus (1997, p. 2) has remarked how a full 1/3 of the companies listed in the 1970 Fortune 500 had vanished by 1983.

[7]"Most people think of complexity in terms of the number of components in a system or the number of combinations one must consider in making a decision. The problem of optimally scheduling an airline's flights and crews is highly complex, but the complexity lies in finding the best solution out of an astronomical number of possibilities. Such needle-in-a-haystack problems have high levels of *combinatorial* complexity (also known as *detail* complexity). *Dynamic* complexity, in contrast, can arise even in simple systems with low combinatorial complexity" (Sterman (2000, p. 21).

[8]Sterman (2002b, pp. 5–7) remarks that dynamic complexity can be detected in those systems which are: (a) constantly changing over time, (b) tightly coupled and interacting, (c) governed by feedback, (d) non-linear, with a changing dominant structure; (e) influenced by history, (f) self-organizing, (g) adaptive, (h) counterintuitive; (i) policy resistant; and (j) characterized by trade-offs.

Framing a dynamic complex problem implies that policy levers are identified to affect system's performance. Performance indicators often portray counterintuitive time behavior.

It also implies the need for identifying trade-offs over time (i.e.: between short and long-term) and across space (i.e.: between a sub-system and another), in relation to alternative policies.

Such trade-offs are due to policy resistance that dynamic complex systems often portray. This phenomenon may imply that, after a given set of policies has been adopted and implemented to fix a problem, the system may respond by showing a performance improvement in the short run. However, in the long run, problems may bounce back, often stronger and more pervasive than in the past.[9]

For instance, unintended consequences underlying trade-offs over time may arise when discretionary cost reduction policies are adopted. In order to fix economic losses, company policy makers may reduce Research & Development (R&D) or training costs. In the short run, such policy may determine an improvement in the income rate; this might allow the firm to plan an increase in discretionary costs for the future (see the balancing loop implicit in the mental models of policy makers, as depicted in Fig. 1.1a). However, this policy often causes a further reduction of financial performance in the long run. In fact, lower R&D or training investments would respectively increase—after a delay—the company product portfolio and human capital obsolescence rates. Figure 1.1a shows the short and long term effects of such policy. In the short term, discretionary cost reduction counterbalances the income shortage. This is expected to allow policy makers to restore discretionary costs, after a delay, to their previous values. However, in the long run, discretionary costs reduction depletes the company intangible assets, which will dampen the perceived quality of products or services. This will further increase economic losses. To counteract this problem, decision makers might be inclined to look again for a symptomatic solution, i.e.: a new reduction in discretionary costs ("snowball" vicious reinforcing feedback loop in Fig. 1.1a).

Figure 1.1b provides a similar example of policy trade-off over time, applied to local government.

In this case, a Municipality implements an indiscriminate reduction of urban renewal investments with the aim to counteract negative trends in its cash flows. Similarly to the previous example, this policy may improve financial performance in the short run. However, in the long run it would reduce quality of life in the city and its suburbs. In fact, a reduction of urban renewal investments might gradually worsen the city infrastructures (e.g. concerning transportation, garbage collection, water

---

[9]"Policy resistance occurs when policy actions trigger feedback from the environment that undermines the policy and at times even exacerbates the original problem. Policy resistance is common in complex systems characterized by many feedback loops with long delays between policy action and result. In such systems, learning is difficult and actors may continually fail to appreciate the full complexity of the systems that they are attempting to influence. Often, the most intuitive policies bring immediate benefits, only to see those benefits undermined gradually through policy resistance" (Ghaffarzadegan et al. 2011, p. 24).

(a)

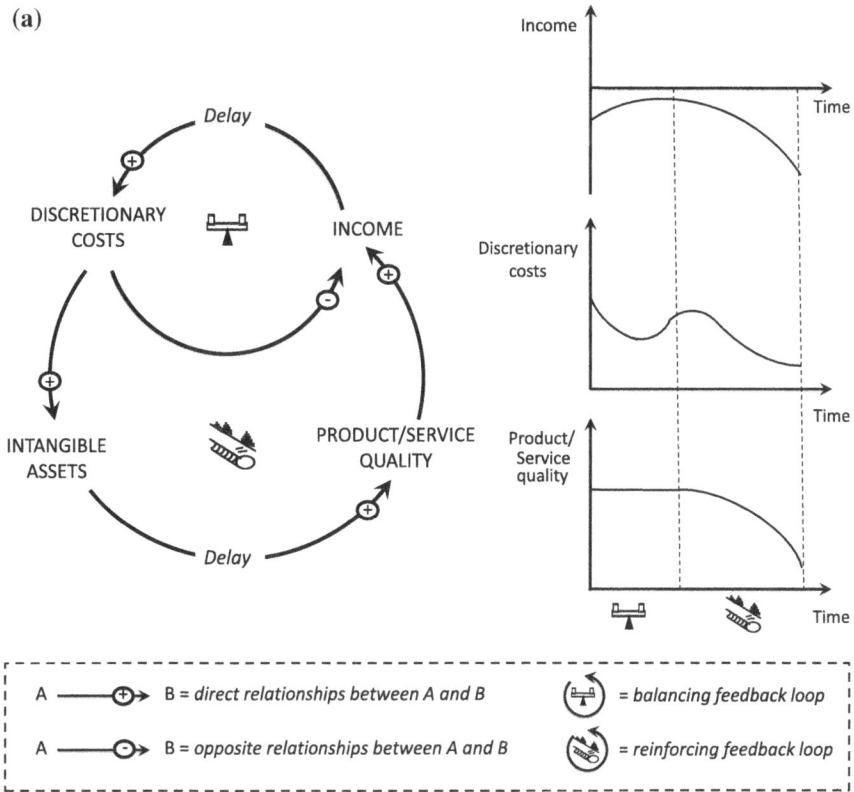

Fig. 1.1 a Unintended effects from indiscriminate discretionary costs cutting. **b** Unintended effects from indiscriminate reduction in urban renewal investments. **c** Unintended effects from overinvestment policies in the commercial sub-system

distribution, parks) and the quality of supplied services. Inertial forces characterize such phenomena: the perception by citizens of the effects that such indiscriminate investment cuts generate on quality of life is not immediate. As Fig. 1.1b portrays, in the short run, reductions in urban renewal investments would not generate any major change in urban quality of life. In fact, perhaps even over a few years after such policy has been adopted, the functionality of city's infrastructures might not be significantly affected by the lack of maintenance. Therefore, the adopted policy might look as a good fix to the financial shortage (balancing loop dominance in systems behavior in Fig. 1.1b). The unintended effects of this policy will be felt by the community, maybe after long (physical and perception) delays, when quality of life will progressively deteriorate due to the loss of functionality of the city's infrastructures.

The outcome of such phenomenon will be a loss of attractiveness of the city and, therefore, a reduction of its population. This will, in turn, affect the potential tax revenues for the municipality. Therefore, in the long run the described "shift in loop dominance" (from a balancing to a reinforcing loop) will further exacerbate financial difficulties.

(b)

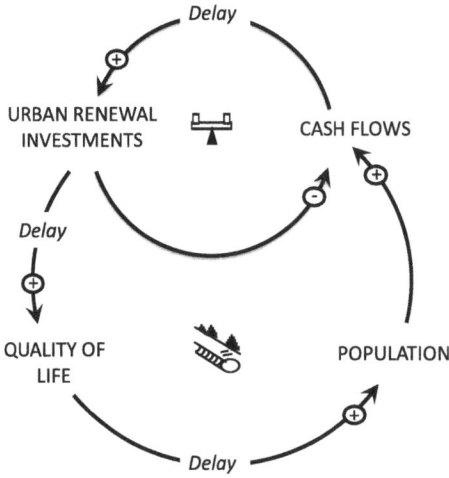

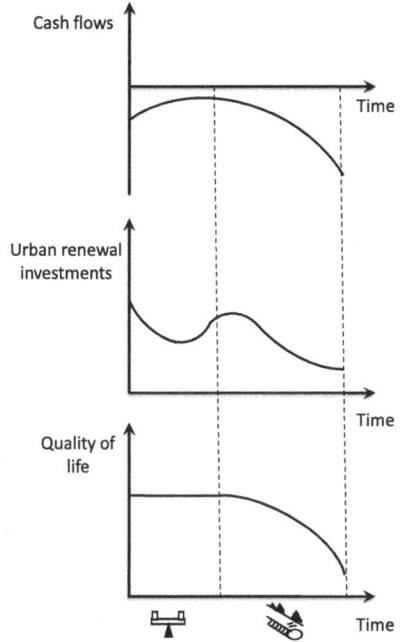

(c)

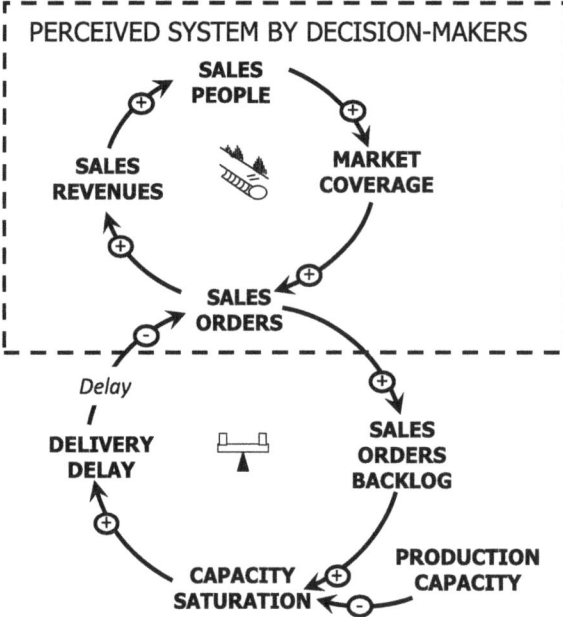

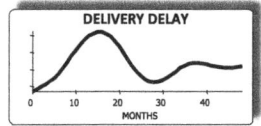

Fig. 1.1 (continued)

An example of policy trade-offs in space (and time), is portrayed in Fig. 1.1c.

In this case, in order to sustain a market growth policy, decision makers tend to overinvest in the commercial system, while they disregard the production system. Imagine a small business entrepreneur whose mental models (because of past experiences and natural aptitudes) are focused on the idea that commercial growth must drive capacity investments. Since the company has an unexploited production capacity, the entrepreneur decides to hire more sales people, with the aim to increase the company customer base, and sales orders. In a competitive system where market coverage (i.e. the ratio between the company sales agents and the potential customers) is a critical success factor, this policy may increase the sales order and revenues rates. With the aim to sustain commercial growth, if the entrepreneur is biased against capacity investments and if full capacity saturation has not been reached yet, the company will reinvest a fraction of the higher revenues in hiring more sales agents.[10]

Though in the short run such policy might be able to foster market growth, in the long run it would prove to be unsustainable, because of the inertial effects caused by an increasing sales orders backlog on capacity saturation and on delivery delay. If delivery delay is another critical success factor in the competitive system, a rise in perceived delivery delays beyond the thresholds tolerated by customers, would reduce the sale orders rate (see balancing loop acting as a limit to market growth, in Fig. 1.1c). Since a reduction in the sales orders rate would contribute to decrease the sales order backlog and capacity saturation, this phenomenon might further strengthen the negative bias of the entrepreneur towards capacity investments.

To the biased entrepreneur, it can be unclear the rationale of customers' behavior, who have reduced their orders, in spite of aggressive commercial strategies.

Such a sectoral perspective implies the risk that a further increase in the sales-force, gives rise to a new sales increase, which could deceive the entrepreneur, who might feel to have the problem fixed. Nevertheless, on a longer time horizon, the higher sales orders would gradually determine a further increase in the production capacity saturation (i.e. the *problem* to fix) and delivery delays (i.e. the *symptom* to promptly and selectively perceive). This would in turn erode the business image, thereby reducing sales again.

If one would analyze the above problem in a static and *post facto* perspective, it might appear as self-explaining that, in order to satisfy a growing demand, it is necessary to anticipate production capacity bottlenecks and to couple salesforce increase with capacity investments. Nevertheless, the need of such policy could not be so evident to a marketing-oriented entrepreneur who is not inclined to bear a higher level of rigidity and economic risk, also due to financial constraints. The entrepreneur's culture and personal traits and his full involvement in the continuous bundle of current management activities might tackle a prompt perception of the slow, although progressive, decline in production and delivery delay performance.

---

[10]The example is borrowed from: Forrester (1968).

Unfortunately, when the weak signals of change become so evident to require a prompt correction of the flow of events, it could be either more difficult (and expensive) than in the past, or too late to restore the lost business image.

As described by Senge (1990)[11] business ventures are often destined to fail, due to—rather than erroneous long-term decisions—a lack of selective perception of weak problems that were becoming day-by-day more and more serious for the firm. Very seldom, pure instinctive and emotional decision-making is likely to support organizational growth. Creativity and "flair for business", are a strength; however, if they are not supported by the awareness of the relevant system in which the organization operates, they are likely to give rise to a waste of resources.

Organizational crises are rarely generated by chance, or fate, as it might appear at first sight to those unsuccessful managers that are used to blame external factors, such as the public sector or macro-economic cycles. Usually, the deep causes of crises are far from being related to sudden and inescapable events. On the contrary, they gradually arise as a product of the concurrent action over time of different variables pertaining to the relevant system.

The relevant system, related to a given problem behavior (Forrester 1961; Richardson and Pugh 1981, pp. 42–43; Sterman 2000, pp. 222–225), does not usually coincide with the internal boundaries of the firm. It also embodies a wider range of variables belonging to other external sub-systems, e.g. related to the competitive and social environment.

1.3 Improving Mental Models and Fostering Strategic Learning Through System Dynamics

It is not easy to detect weaknesses in mental models through only accounting or statistical tools, as they may provide detailed and static pieces of information; they also do not generally provide non-financial information (e.g. delivery delay).

A significant contribution to the analysis of mental models' weaknesses has been given by Argyris and Schon (1978), and Argyris (1985), who described how very often people's behavior in an organization system, is affected by internal inconsistences

---

[11]"Maladaptation to gradually building threats is so pervasive in systems studies of corporate failure that it has given rise to the parable of the 'boiled frog'. If you place a frog in a pot of boiling water, it will immediately try to scramble out. But if you place the frog in a room temperature water, and don't scare him, he'll stay put. Now, if the pot sits on a heat source, and if you gradually turn up the temperature, something very interesting happens. As the temperature rises from 70 to 80 °F, the frog will do nothing. In fact, he will show every sign of enjoying himself. As the temperature gradually increases, the frog will become groggier and groggier, until he is unable to climb out of the pot. Though there is nothing restraining him, the frog will sit there and boil. Why? Because the frog's internal apparatus for sensing threats to survival is geared to sudden changes in his environment, not to slow, gradual changes" Senge (1990, p. 22).

between what a single "actor" asserts (so called espoused theory) and the way he or she behaves (so called theory-in-use), which is affected by mental models.

Also, it has been demonstrated how people are often prone to keep their own ideas, opinions and beliefs unchanged, even though contingent events seem to contradict them. Usually such phenomenon, defined as "cognitive dissonance" (Festinger 1957), is a source of tension and anxiety, on which it is important that learning facilitators are able to act.

Individual and organizational learning is always the outcome of a mental models adaptation (and change) process, which is a result of an "intelligent" analysis of the real world.

When an organization operates in dynamic complex systems, if mental models are only supported by the use of accounting or statistical data, it is likely that decisions will be affected by a conservative approach that will give rise to a "single loop" learning process (Argyris and Schon 1978). Single loop learning implies that decision makers do not question organizational routines for a relatively long period of time. In such context, change mainly occurs through discrete events, which are set by only formal plans. This view portrays strategic management as only a matter of long-term decision-making; a sharp distinction is made between strategic design and implementation. The perils of single loop learning are related to the rigidity of decision-making processes, specifically when organizations operate in systems characterized by a high level of complexity and unpredictability.

In order to overcome accounting and mental models' weaknesses, it is necessary to enhance a "double loop" (Argyris and Schon 1978) learning process, which allows decision makers to evaluate consistencies in their mindsets, i.e. the way they frame problems and strategic issues.

Double loop learning can be fostered by the System Dynamics (SD) approach, in order to:

- make mental models explicit;
- assess mental models' consistency, and
- improve mental models.

SD modeling allows one to implement a double loop learning process facilitated by modelers. On the one side, the modeling process is rooted on the elicitation of decision makers' perceptions of the real world (i.e. the system's structure underlying detected behavior). On the other side, decision makers' mental models are challenged through model validation (i.e. the search for a consistency between the model hypotheses on the system structure and the simulated behavior).

Furthermore, SD modeling challenges mental models through simulation. Once the model has been validated, decision makers are enabled—through the support of a modeler/learning facilitator—to test in a "protected" environment the consistency (e.g., in terms of robust trade-offs perception) of their own decisions.

Therefore, as Fig. 1.2 shows, through a virtual reality (i.e. SD modeling and simulation) decision makers can be helped to perceive the role of the factors shaping their own rationality. According to a "single loop" learning approach, information feedback (e.g. reporting) directly affects decisions, which in turn affect

Fig. 1.2 System dynamics modeling and double loop learning (adapted from Sterman 1994)

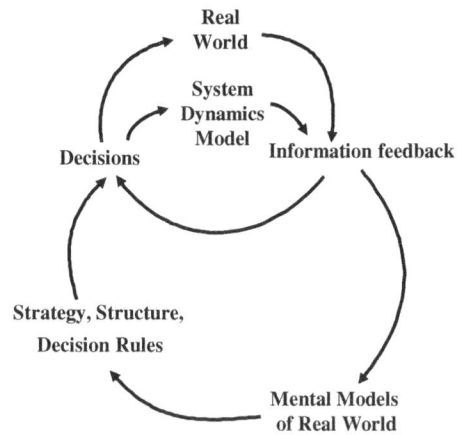

the real world, and the information feedback again. Conversely, through a "double loop" learning approach people may reflect how information is filtered through their mental models, which in turn affect organizational strategy, structure and decision rules, leading to decisions, that will ultimately affect the real world. Detecting and framing the hidden and lower portion of decision-making, as shaped in Fig. 1.2, can significantly improve the quality of performance management.

As remarked by Sterman,[12] policies undertaken in a system often prove to be unsustainable in the long run, because of decision makers' inclination to frame in their minds their past experiences as a sequence of events. For instance, in the attempt to identify a problem to tackle, decision makers may stop their own analysis to superficial and contingent events and corresponding measures, like the following: "inventory is too big", or "sales turnover this month has dropped". To look for the causes of such events—and to counteract them at their own roots—decision makers should be supported by learning facilitators to frame the relevant system underlying the single patterns of dynamic behavior, as shown by accounting or statistical reports. For instance, inventory might be too big because sales have suddenly decreased. And this, in turn, might have been caused by competitors' decision to reduce their sale price. At this point, one might ask why competitors might have dropped their own prices. The causes of competitors' price reduction could be then linked to other phenomena, and so forth.

Therefore, a linear way of thinking may lead policy makers to superficially diagnose a negative gap between budgeted and actual sales and to only identify symptomatic responses to the perceived problem. For instance, this perspective might suggest them to tackle a sales reduction (i.e. the symptom) by reducing prices or by increasing incentives to salesmen, without questioning the problems that are behind the detected symptom and the effects that the intended policy might generate in the long run.

---

[12]Sterman (2000, pp. 10–11).

Fig. 1.3 The need of a feedback approach to perceive policy resistance effects generated by short-term policies in a dynamic complex system (adapted from Sterman 2000, p. 11)

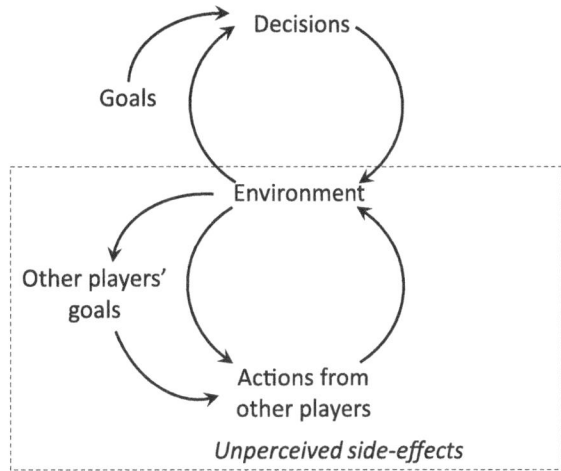

Decisions

Goals

Environment

Other players' goals

Actions from other players

Unperceived side-effects

A more careful analysis should suggest that the actions undertaken by the company decision makers are not external to the relevant system generating the detected symptom. Such policy makers rather belong to a wider system, whose performance is also affected by the decisions of other players (e.g. competitors). Therefore, though in the short run adopted decisions to affect the environment (e.g. customers) might allow the firm to fix the symptom, in the long run they might also contribute to change the policies undertaken by other decision makers in the system. For instance, the sales increase caused by the company policy might generate a reduction in competitors' sales (assuming a constant potential market). With the intent to recover the lost sales volumes, competitors might then respond to their sales reduction by also dropping their own prices or by undertaking other aggressive commercial policies. The latter provides a second example of trade-off in space (and over time) in policy making.

Figure 1.3 illustrates how, in order to avoid the phenomenon of policy resistance, it is necessary to extend the boundaries of the perceived system and to perceive the feedback loops generated over time by the undertaken decisions. Such view of organizational performance can be supported through the SD approach.

By adopting such view, if policy makers operate in a dynamic complex system, and their learning/decision making processes are supported by modelers (or "learning facilitators"), they can be enabled to perceive the relationships between the system feedback structure and its behavior (i.e. its time performance).

1.4 System Dynamics: A Methodology to Foster Organizational Learning and Performance Management

Understanding the causes underlying perceived phenomena, when one acts in dynamic complex systems, does not require that systems complexity be simplified (i.e.: reduced or ignored). It is, rather, needed an approach that enables decision makers to manage (i.e. to explore or deal with) complexity (Morecroft 2002).

This implies that a learning-oriented approach is adopted, i.e. that human capital development policies may enable decision makers to learn and practice systems thinking.

It also requires that planning and control (P&C) systems[13] are designed and used (Bhattacharyya 1971; Chenhall 2003; Das 2011, pp. 22–28; Ferreira and Otley 2009; Flamholtz 1983; Flamholtz et al. 1985; Hofstede 1978, 1980; Maciariello 1984; Ouchi 1979) to enable decision makers to set a range of targets that are compatible with the sustainable development of their organization, and to identify the key variables and policies impacting on them (Otley 1994, p. 297).[14]

Though planning and control are usually described as different functions, if organizations operate in dynamic complex systems (Lorange and Scott Morton 1974) the processes associated with: (1) setting goals and objectives, (2) assigning responsibilities, (3) monitoring results through the use of feedback and feed-forward mechanisms (De Haas and Kleingeld 1999; Nklit 2000), (4) motivating and rewarding decision-makers, and (5) reporting performance cannot be conceived as components of isolated systems. An integrated and systemic view of P&C, rather, provides a sound basis for implementing performance management (Flamholtz 1996; Otley 1999, 2001). In this perspective, particularly in the last two decades, research and practice on P&C have been emphasizing more the purpose (i.e. performance management)—rather than the means—related to such systems. Therefore, the term "performance management system" has been used more intensively than the traditional one (i.e., "planning and control system").

---

[13]The concept of "planning and control system" dates back to Robert Anthony's taxonomy (Anthony 1965). According to Anthony's framework, the P&C system is identified as a set of functions, consisting of three sub-systems (Anthony and Govindarajan 2007, p. 6), i.e.: Strategic planning, management control, and operational control. All these subsystems are described as processes aimed at affecting different, but interrelated, functions. In fact, (1) strategic planning focuses on deciding the objectives of the organization, the resources to be used in attaining them, and the policies for governing the acquisition and use of these resources; (2) management control focuses on how "resources are obtained and used effectively and efficiently in the accomplishment of the organization's objectives"; and (3) operational control focuses on how "specific tasks are carried out effectively and efficiently." Maciariello (1984, p. 3) identifies three main components of a P&C system, i.e.: (1) organization/responsibility centers structure; (2) the performance measures and rewards and the systems of information flows; (3) the process, which consists of setting targets, evaluating performance, making decisions, and getting feedback from them.

[14]A more detailed analysis of this concept will be outlined in the next chapter of this volume.

In this regard, Moynihan (2008, p. 5) defines performance management as "a system that generates performance information through strategic planning and performance management routines and that connects this information to decision venues, where, ideally, the information influences a range of possible decisions".[15]

If related to an organization, the concept of performance can be synthetically referred to the results in the implementation of policies over a given period of time. Performance management should not be conceived as a mechanistic or bureaucratic task. On the contrary, particularly those organizations operating in dynamic and complex contexts, performance management can be enhanced by the use of a SD approach.

The role of "learning facilitators", whose skills should combine SD modeling with performance management, should foster the shift of mind described in this section.

This is the domain of dynamic performance management, which is the topic of this book.

System Dynamics is a methodology that was developed since the 50s at MIT (Cambridge, USA) by Jay Forrester and later spread in many Universities, research centers and other organizations throughout the world.

Since a system is a group of interconnected variables, each of them performing a specific function according to a common and superior goal, organizations can be framed as systems of functional units (e.g. procurement, production, commercial, R&D, finance), whose role is to contribute to business survival and sustainable development, by pursuing excellence in financial, competitive and stakeholder relationships.

In fact, a business is an interconnected dynamic operations system made up of several sub-systems, and interacting with an external environment consisting of different sub-systems too.[16]

For instance, inside an enterprise one can identify the following sub-systems:

- Strategic business areas;
- Processes related to the management of specific resources (e.g. human, financial, etc.);

---

[15]Where referring to system performance, Forrester (1961, p. 116) says: "By 'performance' we mean here such measures of a system as profitability, employment stability, market penetration, product cost, company growth, price fluctuation, investment required, and cash position variations—those characteristics that indicate the 'desirability' of system operation".

[16]As remarked by Coda (2010, p. 17): "The aim of business scholars is to study not so much economic phenomena per se, taken separately from the organized activities in which they occur, but these very activities as they take place in organizations. These activities are essentially management, organization, and accounting. Taken as an interconnected whole, this is what constitutes the object of study for Management Theory, or, put another way, the science of business administration (Zappa 1927)". Zappa (1937, Sect. 4) stated: "An organization, like every economic and regulated unity, is something more than the sum of its components; the complex has properties that its elements do not have and which these elements do not help to define; and the features of the complex cannot be given by a mere composition of the features of these elements either" (Dagnino and Quattrone 2006).

- Processes pertaining to specific functional areas (e.g. Finance, Marketing and Sales, Production, R&D);
- Processes referred to the functioning of given operational mechanisms (e.g. budgeting, rewards, careers).

Furthermore, examples of external sub-systems related to a wider environmental system of a business can be referred to its competitive and stakeholders systems, which could be further partitioned in sub-systems.

SD modeling can provide organizational decision-makers with proper 'lenses' to frame dynamic complexity and to enhance traditional performance management approaches.

Popular approaches to performance management involve spreadsheet models and/or accounting packages. Spreadsheet simulation modeling, based on balance-sheet data extrapolation on a periodical basis, can provide decision-makers a limited support to plan for sustainable growth. In fact, spreadsheet models generally lack flexibility (Shrage 1991): they are usually based on a linear, static, narrow, and unselective approach. Their perspective is linear and static, as it is based on the projection of balance-sheet data and omits to consider feedback loops; it is narrow as it is only focused on financial measures and does not make explicit how external variables (e.g. competitors' policies) affect performance; it is unselective, since—being based on analytical and hierarchical databases, which give rise to detailed reporting—they lack to deliver prompt and relevant information for decision making.

Simplifying systems analysis only apparently allows a reduction in complexity, as complexity and unpredictability ought to be understood and properly handled through modeling. This focuses on:

- Interdependencies between variables;
- Relationships (including non-linear) between policy levers and affected variables;
- Delays.

SD also acts as a "bridge" connecting two broad domains which have been traditionally kept separated, i.e.: the formal quantitative-mathematic approach aimed at finding optimal solutions to business problems, and the concrete management world, which requires a better understanding of complex systems and dealing with unpredictability.

Traditional Operational Research has been used to apply "optimizing" techniques, which support a simplified analysis of reality (Forrester 1961, pp. 1–9). When decisions are made in dynamic and complex systems, such input-output models—which often provide the basis on which Decision Support Systems (D.S.S.) are developed—do not properly allow decision makers to discern cause and effect relationships between relevant variables and to focus on selected policy levers on which to act in order to affect a system's performance. They may allow one to detect optimal solutions (i.e. to identify a set of policies that are supposed to meet the maximum or minimum value of a given target parameter), based on systems' constraints. Through this perspective, the system is seen as a "black box", from which a given output is obtained from a set of inputs. However, at least three problems can be associated with this view, i.e.: (1) a condition that will enable management to convert inputs into the expected output is that the assumptions behind the decision model will come true; (2) the output is not linked to any outcome in the long run (therefore, a static and short term measurement of the effects of decisions is enough to assess performance); (3) no other player, except from organizational decision makers, will be able to significantly affect the results of implemented policies.

On the use of such techniques, a remarkable problem is that when organizations operate in complex and unpredictable systems, it is unlikely that all analytical hypotheses on which an optimization model is based will be empirically verified. When decision makers do not properly frame a system's structure, modeling runs the risk to be no more than a fallacious exercise, which is likely to generate a dangerous illusion of control. In the described decision context, rather than optimizing, it is much more important framing the system's structure and behavior and to learn from a continuous comparison between the real world and the model.

As Forrester (1961, p. 57) remarked: "The mathematical model is not necessarily more 'accurate' than the verbal model, where by accuracy we mean the degree of correspondence to the real world. A mathematical model could 'precisely' represent our verbal description and yet be totally 'inaccurate'". The formalistic approach characterizing rigid quantitative methods, whose main concern is the search for such 'precision', may often lead to omit modeling many intangible variables that strongly affect a system's performance. They are claimed to be intrinsically outside of the domain of modeling, since they cannot be objectively measured. However, Forrester remarks (1961, p. 57) that "to omit such variables is equivalent to saying they have zero effect—probably the only value that is known to be wrong!".

Modeling for learning (De Geus 1994; Lane 1994) ought to be so flexible to allow policy makers to continuously adapt the model to an ever-changing real world. Since facilitated learning, enhanced communication among decision makers, and mental models elicitation are the critical means to improve the quality of planning and performance management, the modeling process (i.e. system's mapping) takes a central role. SD modeling implies that decision makers participate to model building. Their own explicit and tacit knowledge—that Forrester (1994) defines as "mental database"—together with coded data from accounting models and the wider information system in an organization (including external data), are a prerequisite to build a SD model. Facilitated modeling sessions are a vehicle to raise doubts and to foster a strategic dialogue on the hypotheses upon which past decisions have been made, and to suggest how alternatively a given cause-and-effect relationship could be modeled. Therefore, they can support outlining and assessing new hypotheses that may foster a more robust planning and performance management.

In the described context, it would be much more supportive for the quality of decision making and performance management a SD model, that has proven to be 'wrong' after its implementation—but that has been developed through decision-makers' participation—than a theoretically 'perfect' SD model (i.e. able to predict the future) that has been sketched from only outside the organization. The purpose of SD is not the tool (i.e. the model), but the process (i.e. modeling and learning).[17] As remarked by Sterman (1985, p. 521), "Models rarely fail because we used the wrong regression technique or because the model didn't fit the historical data well enough. Models fail because more basic questions about the suitability of the model to the purpose weren't asked, because a narrow boundary cut critical feedbacks, because we kept the assumptions hidden from the clients, or because we failed to include important stakeholders in the process".

SD models, rather than being prescriptive, are descriptive. They help people to better frame the systems in which they operate and to figure out different possible outcomes related to several sets of adopted policies, according to different scenarios.

1.5 Framing Problems into Closed Causal Boundaries Through System Dynamics. The "External" and "Internal" Views as Complementary Perspectives in SD Modeling to Support Decision Making and Performance Management: Implications for the Public Sector

SD models are based on a feedback view of systems, seen as closed boundaries, i.e. embodying variables that explain system behavior. Inside model boundaries those variables that mostly affect the problem being investigated are included, regardless they are internal or external to the organization where decision makers perform the modeling.

The relevant system often includes several decision-making units, operating in different organizations; it is analyzed as a closed loop system.[18] For instance, if the

---

[17]In this regard, Sterman (2002a, p. 521): "Because all models are wrong, we reject the notion that models can be validated in the dictionary definition sense of 'establishing truthfulness', instead focusing on creating models that are useful, on the process of testing, on the ongoing comparison of the model against all data of all types, and on the continual iteration between experiments with the virtual world of the model and experiments in the real world. We argue that focusing on the process of modeling rather than on the results of any particular model speeds learning and leads to better models, better policies, and a greater chance of implementation and system improvement". The same concepts are remarked in Forrester (1985).

[18]"Forrester's use of the term 'closed' means 'causally' closed. His use is different from the notion of a closed system in general systems theory, which refers to a system that is 'materially closed', that is, does not exchange material or information with anything outside the system boundary. Forrester's 'closed boundary' systems are, in general systems theory terms, 'open systems' because they include little clouds representing sources and sinks of material outside the system boundary" (Richardson 2011, p. 241). See also Richardson (1991, p. 298).

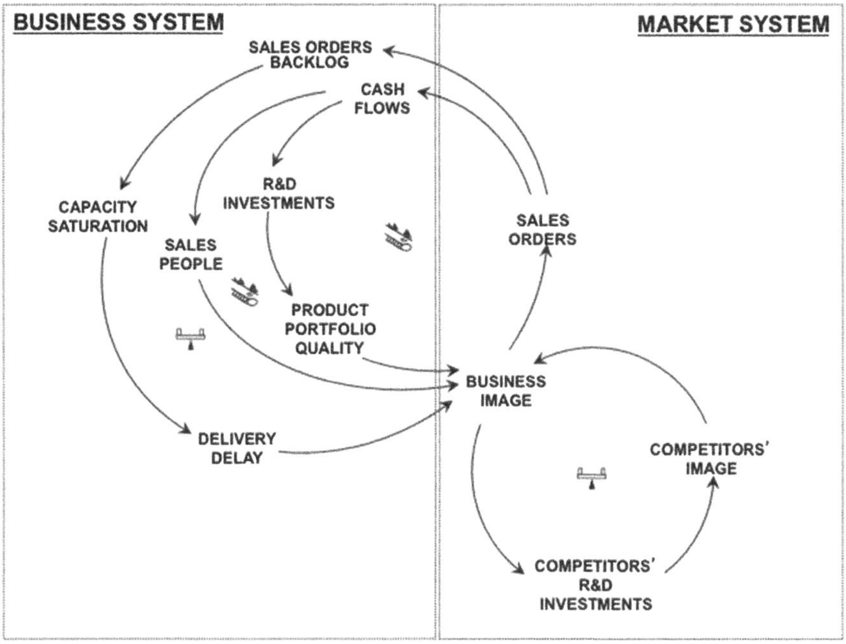

Fig. 1.4 An example of a closed loop system

purpose of modeling is to frame business image accumulation and depletion processes, a SD model might both embody internal variables (in respect to the business
performing the modeling) and external variables, related to the market subsystem.

Figure 1.4 shows an example where business image is not only affected by
feedback loops that can be directly influenced by company policies (i.e. product
portfolio quality, salespeople efforts, and delivery delay). Image can be also affected
by competitors' policies that may counteract its improvement through R&D
investments, aimed to improve competitors' product quality and image. This would
reduce the business relative image (balancing loop).

If one takes the point of view of each decision maker on behalf of whom a SD
model is developed, such perspective could be defined as "external", since it does
not primarily reflect the observation point from which each involved player per-
ceives the system from inside the organization (i.e. the institution) where he or she
operates. In other words, an "external" perspective primarily implies an analysis of
the relevant system per se, rather than that of a specific decision maker.

A clear example of an "external" view in SD modeling is provided by the
seminal work of Jay Forrester, on industrial dynamics (Forrester 1961). In this field,
the main focus of modeling is on the industry (rather than on a single business)
supply chain.

If specifically referred to the public sector, an "external" feedback systems view
suggests very insightful issues. Traditional public sector applications of SD have

been oriented to the study of industries, such as energy (Sterman and Richardson 1985; Davidsen et al. 1990; Ford 1999; Dyner and Larsen 2001), health care (Vennix and Gubbels 1994; Wolstenholme 1999a; Lane and Huseman 2008), housing (Goodman 1989, pp. 309–347), tourism (Honggang 2003), agriculture (Thompson et al. 2007), fishing (Moxnes 2000, 2005), water supply (Martinez Fernandez and Esteve Selma 2004), education (Andersen 1990; Richardson and Lamitie 1989).

Many of these studies have been focused on sustainability issues (e.g. urban dynamics, tourism, ecology, energy) (Forrester 1969, 1970; Meadows et al. 1974, 1992, 2001; Saeed 1996; Sterman 2002c; Fiddaman 2007; Moxnes and Saysel 2009); others have been framing the problems associated with the lack of capacity affecting systems performance (e.g. health care).

Other applications have been oriented to depict the structure and behavior of multi-sectoral economic systems, in order to support public policy makers in understanding how wealth is generated in a State or a Region, and what interdependencies exist between different sectors of the economy (Kopainsky et al. 2009). Among such studies, there are those focusing the topic of poverty and wealth creation in developing and underdeveloped countries.

In addition, topics that are longitudinal to different sectors, and significant on a public policy/management point of view, e.g. crime (Homer 1993; Coyle and Alexander 1997; Stephens et al. 2005; Jaen and Dyner 2008) or terrorism modeling (Grynkewich and Reifel 2006), have been developed.

Also, SD applications to individual public institutions (or parts of them) have been done, such as, for instance, in the cases of hospitals, Universities (Barlas and Diker 2000) and even Courts (Bernstein 1994).

More generally, most SD applications into the public sector tend to be focused on understanding the structure and behaviour of systems that usually embody different players, ranging from public to private ones, from organizations to individuals. The main focus is on the wider system, and policy implications for each player can be taken by the light of the responses that the observed system's behavior is likely to give, as a consequence of changes in its structure.

Though such analysis does not disregard the elicitation of the decision areas that each player is in charge of, it does not primarily focus possible dysfunctions or improvement areas inside each single organization belonging to the broader system. For instance, such issues might be related to responsibility overlaps, unattributed roles, inconsistencies, conflicts and ambiguities in decision-making processes, and their consequences on the governance, management and performance of the observed system. On the other hand, an "external" perspective analysis has the merit to provide a neutral basis to frame cause-and-effect relationships underlying the relevant system's behavior, from a point of view that may go by far beyond— both in time and space—that each of the involved players may take.

For instance, if the modeling goal would be understanding the impact on urban life generated by European Union funded works (e.g. on transportation, education, housing, water procurement and distribution infrastructures), and related accomplishment time delays, then an 'external' perspective would primarily focus on the

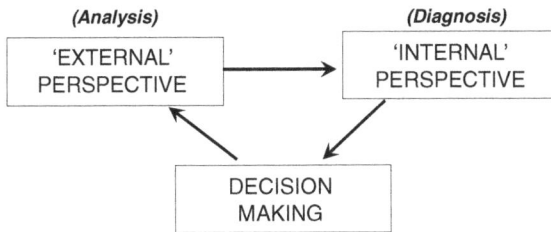

Fig. 1.5 The 'external' and 'internal' perspective as two sequentially-related steps to support public sector decision making and performance improvement through SD modeling

aggregate physical, financial and information stocks and flows associated with projects execution. Such perspective would be likely to adopt a same level of detail and scope in the analysis of factors (e.g. roles, constraints, structure of management processes) impacting on each involved player (e.g. State, Regional and Municipal administrations or other private sector institutions, such as enterprises fulfilling the public works).

A complementary perspective can be defined as "internal". This focuses the wide relevant system by primarily taking the observation point of one of the players (or institution) affecting the system's behavior (Zagonel 2002; Richardson et al. 2004). If such view is adopted, modeling tends to devote a higher level of detail and scope in the analysis of factors that specifically impact on decisions made by the single player (or institution) in the perspective of whom the model is developed. If compared to the "external" perspective, the "internal" one does not necessarily imply the adoption of narrower boundaries for the relevant system in relation to a given problem context. It, rather, implies a more unbalanced or asymmetric analysis, since it tends to focus attention on the way the observed player operates and interacts with the other players within the relevant system.

Both the "external" and "internal" perspectives are necessary in SD modeling. In fact, they complement each other. Although they both support analysis, diagnosis and decision-making processes, the first one seems to be more suitable for analysis, while the second is likely to better foster diagnosis and decision making in specific areas or domains.

Therefore, the two perspectives could be seen as sequential steps in SD modeling to support decision-making and performance improvement (Fig. 1.5).[19]

---

[19]It is worth remarking that the distinction between the above two perspectives relates to the wider methodological discussion in the SD field about the inductive versus deductive nature of modeling. In this regard, it is possible to observe that—though both the external and internal perspective underlie a mix of inductive and deductive approach—the second approach is more ingrained into the external perspective, while the first approach tends to prevail if one moves towards the internal perspective. So, the two perspectives can be seen over a continuum where each of them contributes under a different viewpoint, towards the pursuit of a deeper learning process and better support of decision making of the actors involved with different roles in public sector dynamic complex systems. A similar kind of reasoning also applies if one considers the other important debate in the

Today, an outcome-based performance management approach tends to attract the interest of communities, professionals, and decision makers in both private and public sector organizations. Particularly, if referred to the public sector, it implies a rising attention towards the issues of value generation (Alford and O' Flynn 2009; Borgonovi 2001; Moore 1995; Mussari and Ruggiero 2010), user satisfaction (Osborne and Gaebler 1993), joined-up government (Christensen and Laegreid 2007), and accountability not only on short, but also on long-term results.

In this perspective, traditional performance management practices seem to have fallen short of expectations. Such practices are essentially based on financial models and static reporting; they are focused on an organizational sphere, implying a lack of linkage between outputs and outcomes. They also imply that achieved results are primarily assessed in relation to the effects produced by decision makers on their own institutions.

The lack of framing linkages between outputs and outcomes in performance measurement can lead to a breakdown in the coordination between elected officials and administrators and between different independent agencies, and generate a fracture between managerial and political control. They may also tend to encourage a sectoral and departmental view of administration.

Setting outcome performance measures in an organizational context (e.g. a Municipality) often entails difficulties in framing the effects of the activities carried out by a single agency on the outer local system. In fact, public sector organizations are tightly connected to other independent (loosely coupled) public and private sector institutions, whose policies may affect (positively or adversely) the implementation of organizational programs (Bianchi 2010).

Quite seldom is the area of influence and authority of a given 'actor' operating in a public institution likely to cover the domain of the overall relevant system. The outline of policies and undertaking of management decisions in the public sector is often fragmented through different institutions. Therefore, to affect the behavior of specific decision makers, it is necessary to find a proper balance between the need to model the relevant system in relation to the main problem(s) in the observed sector, and the need to calibrate the analysis on what decision makers are enabled to affect, i.e. by focusing the impact that the specific player could make on the system, in his or her own perspective.

By using a combined "external" and "internal" view, SD modeling can play a crucial role to enrich performance management, and to foster a common shared view of the relevant system's structure and behavior among stakeholders in local strategic planning and execution.

This perspective implies that—for each organization unit—SD models foster a better understanding of the relationships between performance and responsibility.

(Footnote 19 continued)

SD and sociological literature regarding the 'structure vs. agency' relationship. In this regard, the 'social structure' view appears more ingrained into the external perspective, while the 'human agency' view better reflects the internal perspective viewpoint (Lane 2001; Grossler 2004, 2008; Schwaninger and Groesser 2008).

This entails the identification of a self-consistent system of goals and objectives for each unit. Such goals should be embodied in formal plans, together with the definition of the activities that each unit will undertake, in order to build a resource system providing a suitable basis to affect results in the short, medium and long term (outcomes).

Framing problems through this perspective may foster a clear statement and a better understanding of the goals and objectives embodied in the planning documents, and may support managers to figure out how internal processes carried out by back office units should contribute to performance outputs and outcomes.

Such a deeper level of knowledge and awareness that this modeling view provides also prevents dysfunctional behavior consequences that may happen when formal P&C and performance evaluation systems are adopted. Among them: setting easy-to-reach objectives,[20] focusing attention on only a restricted number of objectives to the detriment of others (related to the same goal), identifying the mere accomplishment of activities as targets to reach, bounding objectives to only process or output measures, and confusing means with ends.[21] This behavior is often due to both cultural reasons and lack of proper methods and tools that may foster learning and a change in organizational culture (Busco and Riccaboni 2010) towards the use of performance management as a means to frame dynamic complexity.

Performance management systems designers, policy-makers and managers must be aware of such mismatches in order to avoid a ritualistic or superficial adoption of budgeting and performance measurement, leading to an illusion of control and opportunistic behavior. Conversely, broadening the observed system's boundaries can support a shift of focus from measurement to management, from data collection to systematic use of information, from an input or output to an outcome view of organizational results (Matheson et al. 1997; Moynihan 2008), thereby using performance management systems as a learning tool (Moynihan and Landuyt 2009; Vakkuri and Meklin 2006, pp. 240–242).

Figures 1.6 and 1.7 illustrate how a lack of awareness of the mismatch between the system boundaries perceived at an organizational (or agency) level and those of the broader system generating the outcomes can be a major cause of dysfunctional behavior intrinsic to performance management systems (Bianchi and Williams 2015).

---

[20]This often implies the constitution of slack resources.

[21]On this regard, Flamholtz (1996) remarks the following: "Blau and Scott (1962) reported a study of a public agency whose major goal was to serve workers seeking employment and employers seeking workers. The tasks to be performed included interviewing applicants, helping them to complete application forms, counseling them, and referring them to jobs. To control the interviewers, the agency monitored the number of interviews conducted. The effect of this control system was to motivate the interviewers. They paid attention to the instrumental goals (numbers of interviews), while neglecting the overall (but unmeasured) goal of placing people in jobs". Such practices have been also defined as *gaming*, i.e. a "deliberate manipulation of behavior to secure strategic advantage" (Smith 1995, p. 298). Van Dooren et al. (2010, pp. 162–165) have qualified similar phenomena as "measure fixation" and "cream skimming".

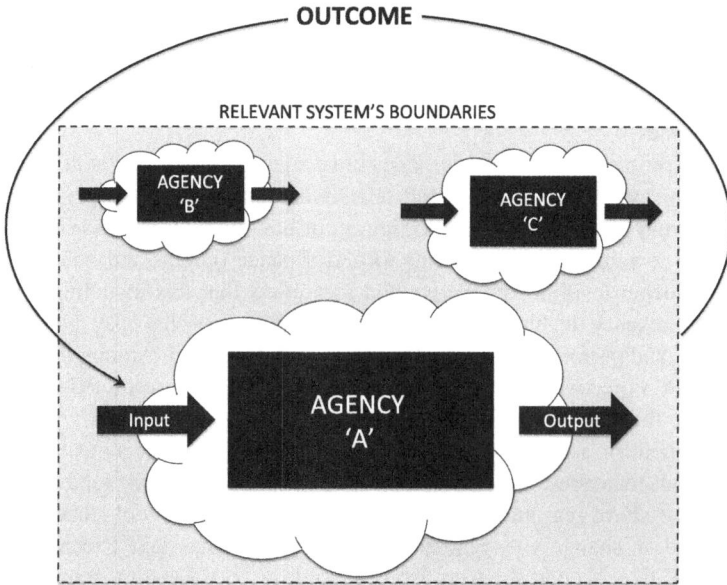

Fig. 1.6 The mismatch between the relevant system's boundaries and the boundaries of each agency's performance management system

For example, in order to be effective in the long run, crime prevention requires collaboration and policy coordination among different institutions. In fact, the effectiveness of a police department is, in the long run, subordinate to its capability not only to prevent or repress crime, but also to the capability of the wider system (schools, courts, other public safety institutions, non-profit sector organizations) to keep crime under control by reducing new and reiterated crime inflow and increasing repressed and solved crime outflows.

As Fig. 1.7 shows, both prevention and suppression policies are relevant leverage points to sustainable crime control. For example, if the police only focus on dealing with uncontrolled—i.e., unsolved—crimes, and were made accountable to only an output performance measure such as the number of crimes solved, even attaining a target for this measure alone may not lead to a reduction in the stock of uncontrolled crime in the region, which is an outcome. Though an increasing pattern of solved crimes might signal police efficiency, it would not necessarily imply that the police are effective. The figure also shows that crime is kept in control through the "B1" and "B2" balancing loops, which enhance effective crime prevention and suppression policies, leading to a steady reduction of the crime level over time. Without synergy between the different actors involved in crime control policy-making, there is a risk that the reinforcing loop "R" originated by repeated crime would prevail over the balancing loops.

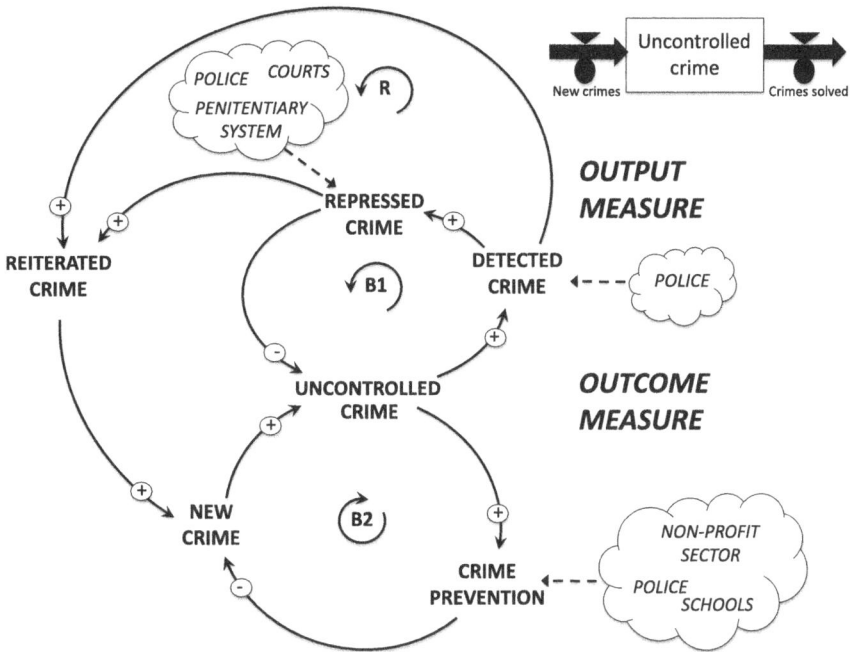

Fig. 1.7 Combining an "external" with an "internal" view to foster sustainable crime control policies by police (from: Bianchi and Williams 2015, p. 402)

Based on the analysis conducted in this section, to move toward this direction it can be useful to combine an "external" view (which is primarily outcome-oriented) with an "internal" view (which is primarily output-oriented) in the design of system dynamics model-based performance management systems (Bianchi 2010, pp. 373–375).

1.6 Conceptual and Simulation Stock-and-Flow Models. Insight Models

Like any other modeling approach supporting decision-making in organizations, SD aims to provide decision-makers with a lens to frame the processes affecting organizational performance, and to set robust policies in the long run.

This can be done through both conceptual (or qualitative) and stock-and-flow simulation (or quantitative) models (Wolstenholme 1999b).

Stock-and-flow models use four main kinds of variables, i.e.: (1) stocks, (2) flows; (3) inputs, and (4) auxiliaries (Fig. 1.8).

In a SD model the resources affecting a system's performance are depicted as level variables, or stocks. The rate at which they change over time is modeled through flow variables.

Fig. 1.8 Main kinds of
variables in system dynamics
modeling

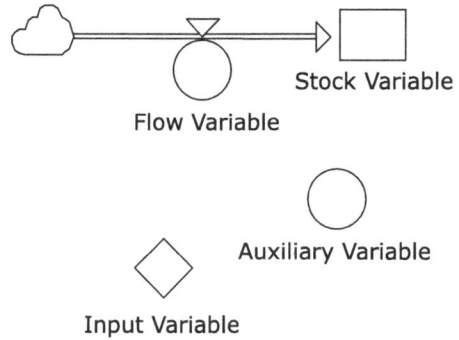

Flow Variable   Stock Variable

Auxiliary Variable

Input Variable

Levels are pieces of information concerning system's conditions (stocks) at a given time; they are a result of an accumulation process triggered by rates. Levels cannot be directly affected, as they represent currently available resources. They can be changed only through rate variables.

Rate variables measure the in-and-outflows affecting corresponding levels in a given time span. They are the effect of management decisions or of external factors impacting on organizations' stocks (e.g. machinery obsolescence, personnel attrition). Flow variables can also measure system's results, which can be affected through the deployment of resources on whose accumulation and depletion processes organizational policies aim to generate an impact.

Those results that are measured by flow variables may influence system's outcomes, i.e. the dynamics of those resources whose corresponding flows cannot be directly affected by the exploitation of other organizational resources. For instance, the flow of new customers accumulating into a customer base (stock) can be affected by the capability of a business to meet delivery deadlines, and therefore to afford order shipping rates (outputs). While sales order shipping rates can be directly influenced through capacity building, the delivery delay (i.e. the ratio between the sales orders backlog and the sales orders shipping rate) and the customer acquisition rate can be only affected by another (short-term, or output) result, such as the shipping rate.

Auxiliary variables can be used to measure those intermediate results affecting output or outcome results measured by flows. For instance, in the example that has been discussed, both the delivery delay and the delivery delay ratio (i.e. the organization delivery delay divided by the delivery delay desired by the market) are measured though an auxiliary variable. Another example where auxiliary variables can be used to gauge intermediate results is the price ratio (i.e. company price/reference price), which may affect the sales orders rate.

Moreover, input parameters represent external constraints or even policy levers on which key-actors may operate, in order to affect—through rates—levels.

Figure 1.9 shows an example of a stock-and-flow model, which frames capacity acquisition policies aimed to affect a shipping rate that is compatible with a delivery

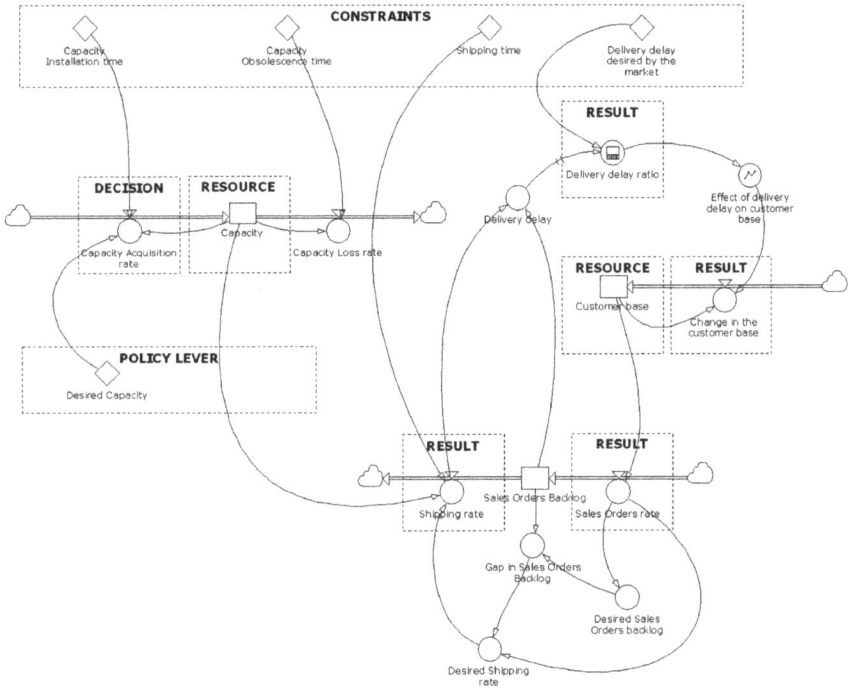

Fig. 1.9 An example of stock-and-flow SD model including: results (intermediate results, outputs and outcomes), resources, decisions, policy levers, and constraints

delay desired by the market. This simple model shows external constraints as input parameters. The model boundaries and time horizon imply that capacity installation and obsolescence time, shipping time and delivery delay desired by the market, are considered as factors that decision makers cannot affect.

The same 'diamond-shaped' symbol is used to model policy levers, i.e. the criteria based on which people aim to affect their resource endowment and deployment, to influence system's outputs, and (through them) outcomes. In this example, the policy lever is the desired capacity level.

Based on the discrepancy between the desired and current capacity level, the model depicts capacity acquisition rate as a valve triggered by a decision affecting the resource, which in turn affects the shipping rate (i.e. the minimum between capacity divided by the shipping time, and the desired shipping rate). The shipping rate is a result since: (1) it can be affected through the resources upon which organizational decisions and policies converge, and (2) it can in turn affect other results over a longer time horizon. In fact, the shipping rate affects the delivery delay, which in turn affects the customer acquisition rate, and therefore it both affects the customer base (resource) and—through it—the sales orders rate.

Through the described perspective, SD stock-and-flow modeling can support the planning process and performance management, since it allows policy makers to track outcome measures, and backwards to identify the outputs and intermediate results affecting them. Their identification allows policy makers to further proceed backwards, by outlining: (1) the resources affecting performance targets, (2) policy levers, (3) constraints, and (4) decisions.

Conceptual models frame the investigated system's structure as shown in the examples portrayed in Fig. 1.1a–c, previously commented. Such models aim to depict the feedback loops explaining system's behavior, with no quantitative data or simulation. For experienced system dynamicists such models may also allow mental simulation, i.e. the expected time behavior of depicted variables can be manually sketched as shown in the previously commented charts. A shortcoming of such approach is that if a conceptual model embodies even more than two feedback loops, the reliability of mental simulation can be bounded. Even for experienced SD modelers it can be highly problematic to mentally simulate the system's behavior as an effect of feedback loop dominance and of different initial states of the investigated system (e.g. in terms of initial values of resources, delays, or external constraints).[22]

Although such kind of mapping has the advantage to be relatively simple and time effective, its limitations imply that, when a more pervasive analysis is required, qualitative modeling needs to be supported by simulation through stock-and-flow modeling.

On the other hand, quantitative stock-and-flow modeling implies a level of detail, an accuracy, and an extension of explored system boundaries which is usually broader than in a qualitative analysis: this is at the same time an advantage and a potential shortcoming, due to the time and costs that developing such models generally implies.

To overcome these potential problems, and even frequently to 'prepare the field' for a more extensive quantitative modeling in a later stage of introduction of SD to support performance management in an organization which is new to this method, a consolidated practice is to develop very simple—or 'insight'—simulation models or even conceptual SD models depicting simplified stock-and-flow feedback structures.

The reason for developing such simplified models—rather than detailed ones— is not only due to the time and costs that large SD quantitative models may imply. It is also related to capacity and cultural issues. Since using the SD method implies a shift of mind in problem framing and decision-making, it may take time for an organization to learn how to use the SD method to enhance performance management systems. They may not have specific SD skills. Particularly small-medium enterprises (SMEs) may have difficulties in providing reliable non-financial and external data to feed the modeling process (Bianchi 2002, p. 322).

The concept of 'insight' (or 'policy-based') model has been introduced by Lyneis (1999). It has been argued (Arthur and Winch 1999) that if the real test of a

---

[22]"People cannot simulate mentally even the simplest possible feedback system" (Sterman 2000, p. 29).

model's validity is based on its 'usefulness' to users (i.e., how effective they found it in improving their understanding and stimulating their thinking), rather than on how closely it mimicked reality of historic time series, then relatively simple models might be valid, and just as effective as highly detailed models (Bianchi and Winch 2006). If simple models can be useful and, particularly, if they could be tailored to an individual firm by a simple and inexpensive process, then simulators could be provided to individual firms to support their particular change management issues. Insight SD modeling is now an established practice that can be used to inform the understanding of processes and is highly dependent on graphic demonstration (Warren 2000, 2008; Winch and Joyce 2006; Wolstenholme 1999b).

1.7   Structure and Behavior Feedback Analysis

Modeling feedback loops triggering organizational performance allows one to map system structure, to capture and communicate the behavior driving processes and the quantification of the relationships to produce a set of equations that form the basis for simulating possible system behaviors over time.

The principle is that, if a system's structure determines systems behavior (Davidsen 1991; Richardson and Pugh 1981; Sterman 2000, pp. 28–29)—i.e. performance—then the key to developing sustainable strategies to improve performance is acknowledging the relationship between structure and behavior and managing the leverage points (Ghaffarzadegan et al. 2011).

The advantage of using this approach is that it places the measures portrayed by the accounting models and embodied in the performance management cycle, within the broader context of the system (Bianchi and Rivenbark 2014), responding to the reality that even simple policy and process changes to impact specific outputs and outcomes are not likely to be that "simple" in organizations (Bianchi et al. 2008).

The SD method allows one to carry out a structure-and-behaviour analysis based on which the reinforcing loops underlying growth can be identified and fostered by proper development policies. Also, reinforcing loops can be associated with corresponding balancing loops, which provide a source of limit to the growth or stability in the investigated system. By promptly detecting and counteracting balancing loops, decision makers can foster sustainable development.[23]

A reinforcing (or positive) loop generates an exponential behavior in the key variables, due to a virtuous or a vicious circle, leading to a growth or crisis process. Two examples are portrayed in Fig. 1.10a, b, which combine a simple stock-and-flow with a conceptual model (i.e. influence diagram) of customer

---

[23]Loop polarity is detected by multiplying the algebraic signs of each arrow, symbolizing a direct or opposite relationship between variables. A reinforcing loop is determined by a positive polarity, while a negative polarity implies a balancing loop. While a reinforcing loop generates multiplicative growth in the affected variable's behavior mode and implies instability in the system, a balancing loop fosters a stable behavior.

(a) CUSTOMERS

NEW CUSTOMERS

(b)   BANK ACCOUNTS

INTEREST

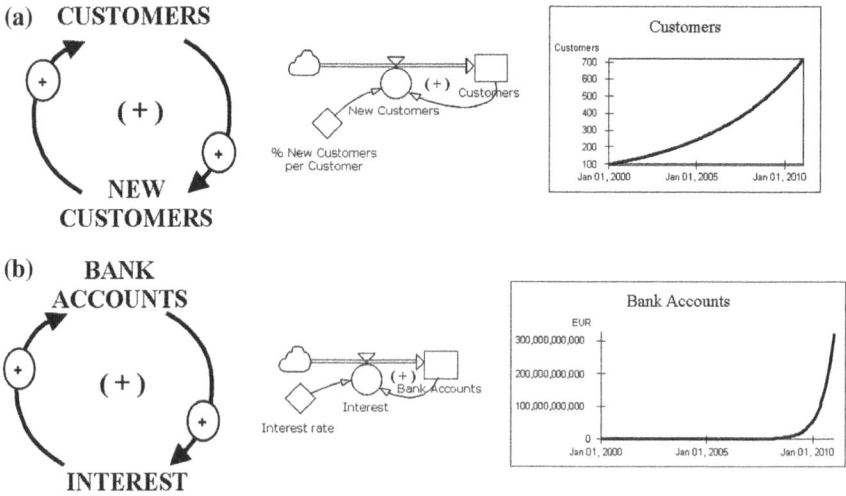

Fig. 1.10 a Reinforcing loop underlying word-of-mouth in customer acquisition. **b** Reinforcing loop underlying accumulated interest on bank accounts

acquisition and interest accumulation processes, respectively. They also combine the two model structures with a time graph showing an exponential growth in the stocks.

A balancing loop may either underlie a draining or an adjusting process, in respect to the affected resource.

Figure 1.11a, b show a draining and an adjusting process affecting the stock of personnel, respectively.

Figure 1.11c combines in a same stock-and-flow diagram the two previous loops. It shows that the adjusting process is dominant over the draining process.

Analyzing feedback loops in a system allows one to investigate on the sources of its behavior, as a function of policies adopted by decision makers and of external factors.

In a later development of the model, constraints can be converted into auxiliary variables, if decision makers' policies can affect them. This implies an extension of model boundaries and often of its time horizon.

For instance, if we consider the last example, the personnel normal attrition time is an input parameter (18 months). Regarding this, decision makers' policies could be oriented to increase such time, e.g. by adopting programs based on incentives, career plans, and training.

An example of how the model portrayed in Fig. 1.11c could be extended to capture such policies is shown in Fig. 1.12. This figure shows how the attrition time is now a function of the normal attrition time and of a multiplier (effect of investment policies on personnel attrition time). This is a normalized function that is modeled through a graph variable identifying several possible multipliers (e.g. between 0.5 and 1.5) in respect to the ratio between the average investment in

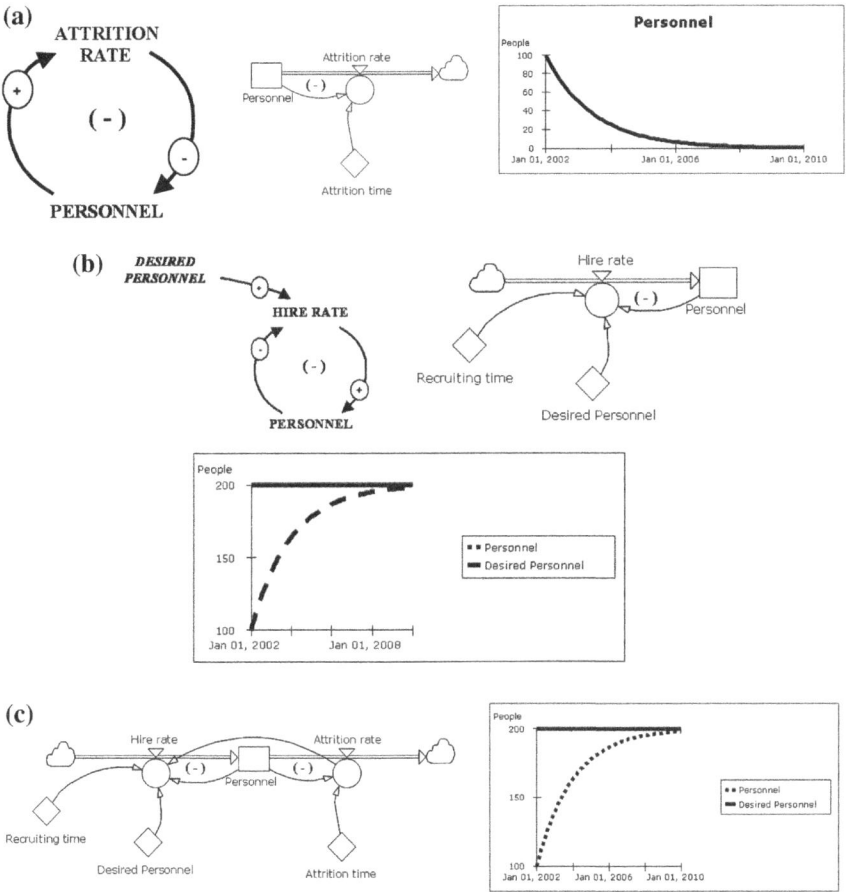

Fig. 1.11 a Balancing loop associated with personnel attrition rate (resource draining). **b** Balancing loop associated with recruiting rate (resource adjustment). **c** Balancing loops associated with recruiting and attrition rates (resource adjustment and draining)

personnel policies (resource) and the normal investment. The higher this ratio is, the higher the multiplier will be, leading to a longer actual personnel attrition time.

The analysis developed so far has illustrated how an SD model is based on explicit statements of policies underlying the decision making process, according to conditions (information on levels, time delays and external input constraints) arising within the system. In accordance with the systems feedback view, management (i.e. decision-making) is seen as a continuous process of converting information into signals, which feed actions oriented to change levels, i.e. to affect resources (Forrester 1961, p. 93, 1994). Such resources will in turn affect the system's performance, that will change the endowment of those resources whose levels cannot be directly influenced by decision makers.

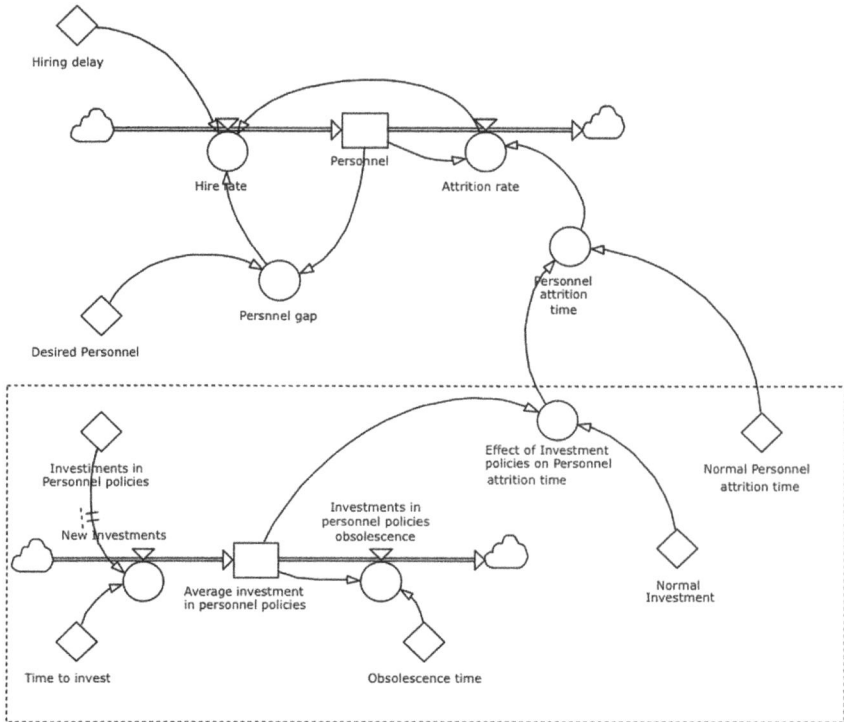

Fig. 1.12 Extending model boundaries

Such conversion process describing management is not always clear and explicit in organizations; it may often involve many decision makers. Making decision processes more explicit through SD modeling and improving them over time may substantially help people to better understand the real world around them and helps them achieve a common shared view.

The quality of policy making depends on information taken into consideration by key-actors and on the way they convert it into action, according to a given set of explicit or implicit "rules". Concerning this, it has been observed that decision-making problems are not usually related to a lack of information (Forrester 1971), but to a proper selection of it, which depends on the quality of mental and information support models.

According to this perspective, decision-making can also be framed as a process of filtering different sets of internal and external pieces of information, related to level and input variables (Fig. 1.13). Such organizational and cognitive filters can be represented as concentric circles: the more they are positioned at the core of decision function, the more they affect it. It follows that, of the many pieces of information directed to a given decision-making unit, only a few of them reach its core.

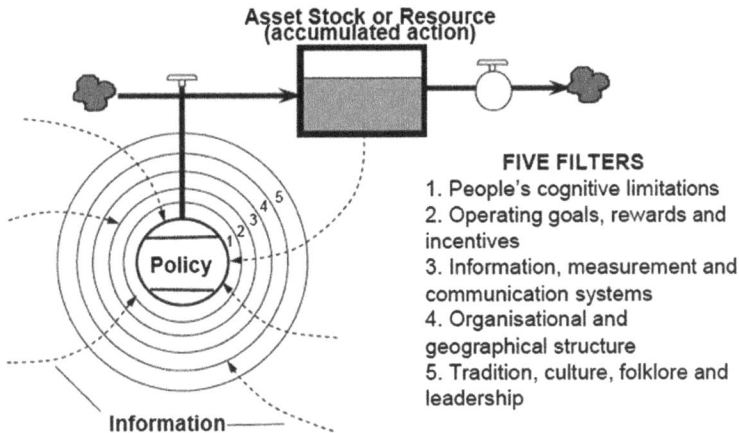

Fig. 1.13 Embodying decision-making and information filters in stock-and-flow modeling (from Morecroft 1988, p. 306)

1.8 System Dynamics Models in the Broad Context of Models Supporting Organizational Decision-Making Through the Planning and Control Function

In order to better focus how SD modeling fits in performance management, it can be useful to position SD into a broader set of models and tools that an information system may adopt to implement the P&C function.

To this end, a conceptual framework borrowed from the taxonomies developed by Anthony, Simon and Keen and Scott Morton will be used.

Anthony (1965) framed managerial activities into three categories, which are intrinsically different from each other, and therefore require distinctive (though interconnected) P&C systems. Such activities are: strategic planning, management control, and operational control.

Strategic planning is "The process of deciding on objectives of the organization, on changes in these objectives, on the resources used to attain these objectives, and on the politics that are to govern acquisition, use, and disposition of these resources" (Anthony 1965, p. 24).

Lorange et al. (1986, p. 10) focused on the same concept by using the term "strategic control". They defined it as "a system to support managers in assessing the relevance of the organization's strategy to its progress in the accomplishment of its goals, and when discrepancies exist, to support areas needing attention". They also remarked that though different terms (and often separate departments) are used to manage (strategic) planning and control, keeping the two activities disconnected is a major cause of failure in the management systems of many organizations. Concerning this, they warned: "That they are two sides of the same coin can be highlighted by the following argument. If we do not control against our plans, we will never know if they have been achieved. Similarly, if we control only against a fixed, rigid set of objectives, we can miss the need to make the changes that mean survival".

In this book, the term "strategic control" will be adopted, to mean a management process combining in a consistent system the broad and interconnected strategic planning and control functions.[24]

The second category identified by Anthony is management control, i.e.: "the process by which managers assure that resources are obtained and used effectively and efficiently in the accomplishment of the organization's objectives" (Anthony 1965, p. 27). Concerning the terms "control" and "controller", Anthony remarked that—though in their ordinary sense they may have unfortunate connotations—a controller is not who should exercise control. People who have the title of controller "should construct and operate a system through which *management* exercises control" (Anthony 1965, p. 28). This role usually relates to a staff unit in the organization chart. It should primarily support a strategic dialogue (De Haas and Kleingeld 1999; Matheson et al. 1997) between the political and managerial level, and the implementation of strategies in an organization. Again, though strategic and management control are conceptually different activities, they need to be consistent each other. The stronger dynamic complexity, the more strategic and management control need interaction, to cascade strategic goals into the organization and to enable management perceive weak signals of change through communication and coordination. Such strategic conversation (Nilsson et al. 2011; Van der Heijden 2004) is fundamental for both designing and implementing policies. Ignoring this approach may generate 'administrative schizophrenia'. In fact, the setting of managerial objectives, actions and targets should imply a deep understanding and communication of the strategies outlined by the political level. On the other hand, the design and assessment of policies cannot ignore the emerging problems and opportunities that can be better perceived on a managerial level (Boyle 1999).

Therefore, outlining robust strategic plans and budgets requires that a controller or P&C analysts facilitate the planning process with feedback (i.e. diagnostic—or *momentum*—control) and feed-forward action (i.e., interactive—or *strategic leap*—control).

Regarding the controller's role in modern organizations, Otley (1994, p. 290) made a very clear statement: "although Anthony had been very specific in suggesting that the basic source discipline for the study of control should be the behavioral sciences, this very rapidly become narrowed to accounting, albeit with a behavioral flavor". He clearly made the point that the control function should go

---

[24]Simons (2000, pp. 303–304) adopted an even broader concept of strategic control, in consideration of the dynamic complexity that organizations are often expected to face. He wrote: "Strategic control is not achieved through new and unique performance measurement and control systems, but through belief systems, boundary systems, diagnostic control systems, and interactive control systems working together to control both the implementation of intended strategies and the formation of emergent strategies".

beyond a focus on only management accounting practices and foster coordination between strategy design and implementation (Kaplan and Norton 2008). This is due to the need for survival by adaptation that organizations should perceive, particularly today in times of turbulence and unpredictability.[25] "The controller is no longer embodied in the higher reaches of the organization; the control function now needs to be embedded at all levels. Only in such a way can the contemporary organization survive in its rapidly changing environment" (Otley 1994, p. 298).

Also, Otley remarked how a more sophisticated use of accounting and other control information is necessary. In this regard, he advocated the need to focus attention not only on (financial) flows, but also on the stocks from which their future dynamics will be affected.[26] This is a key issue remarking how dynamic performance management may enhance the ability of P&C systems to support organizations to deal with complexity and unpredictability.

"The true 'learning organization' is concerned to enhance its overall capability to respond to whatever the external world throws its way ... So a capacity for flexibility is also necessary, the type and amount of which can be investigated using the techniques of simulation and scenario planning" (Otley 1994, p. 297).

Also Lorange et al. (1980, pp. 108–112) emphasized the role of SD modeling to enhance what they define as "strategic leap control", i.e. when "the challenge is to reset the trajectory of the strategy as well as to decide on the relative levels of thrust and momentum for the new strategic direction ... This situation involves a mental leap to define the new rules and to cope with any emerging new environmental factors" (Lorange et al. 1980, p. 11).[27]

The third category of organizational activities identified by Anthony is operational control, i.e.: "the process of assuring that specific tasks are carried out effectively and efficiently" (Anthony 1965, p. 18).

In order to further frame how SD can be positioned in respect to the broad set of models/tools that can enhance the implementation of P&C systems, the commented framework can be matched with H. Simon's taxonomy on organizational decisions.

Simon (1960) distinguished programmed from non-programmed decisions.[28] According to Simon, for non-programmed decisions there are "no specific procedures to deal with situations like the one at hand".

---

[25]"Only those organizations which match their capabilities to the changing needs of the market place and which meet the requirements of other stakeholders will survive in the long-term" (Otley 1994, p. 296).

[26]"This might involve items as diverse as the morale of the workforce, the skill base it possesses, the competitive posture and the capability of the organizational unit, and its potential for adaptation in the light of possible competitive challenges. In such ways it might be possible to counter the bias towards short-termism that seems to exist where financial flows are stressed" (Otley 1994, p. 297).

[27]Also Kaplan and Norton (2001, pp. 311–313) advocated the benefits of SD modeling to foster performance management.

[28]"Decisions are programmed to the extent that they are repetitive and routine, to the extent that a definite procedure has been worked out for handling them so that they don't have to be treated de novo each time they occur ... Decisions are non-programmed to the extent that they are novel,

To describe how information systems can contribute to decision making, in respect to the level of complexity of organizational tasks, Keen and Scott Morton (1978, pp. 86–88) built on Simon's taxonomy. They proposed a slightly different framework by distinguishing: structured, from unstructured and semistructured decisions/tasks. As corresponding terms to Simon's "programmed" and "non-programmed" decisions, they identified *structured* and *unstructured* tasks, respectively. "The degree of potential structure in the task predefines the procedures, types of computation and analysis, and the information to be used. In a highly unstructured task … the decision maker must often rely on personal judgment, especially in identifying exactly what the problem is. By contrast, much if not all of the decision process in a structured task can be automated". Therefore, they argue, to support the fulfillment of structured tasks, "the system will be designed to provide far more fixed routines and sequences of analysis, and to give *answers*" (Keen and Scott Morton 1978, p. 86). Most structured decisions are carried out through operational control, while most unstructured decisions are carried out through strategic planning and control.

Keen and Scott Morton remarked how the distinction between structured and unstructured decisions/tasks is not straightforward. In fact, organizational decisions should be positioned over a *continuum*, over which one may define as "fully structured" or as "fully unstructured" a relatively bounded set of tasks. In fact, also in operational control there can be a component of decision-making implying discretionary subjective judgment.[29] Likewise, also strategic planning and control may entail a—though limited—use of procedures.[30]

Therefore, Keen and Scott Morton identified *semi-structured* decisions/tasks as a third possible typology in their taxonomy. "These are decisions where managerial judgment alone will not be adequate perhaps because of the size of the problem or the computational complexity and precision needed to solve it. On the other hand, the model or data alone are also inadequate because the solution involves some judgment and subjective analysis" (Keen and Scott Morton 1978, p. 86).

If we match the type of decisions that an information system supports (as defined by Keen and Scott Morton) with the P&C level (as defined by Anthony) that decision-making implies, we may further extend the scope of our framework to two more viewpoints, i.e.: (1) the focus of modeling, and (2) the modeling approach.

The focus of modeling is the primary purpose for which models/tools are used to support the P&C function. The modeling approach describes the perspective and the characteristics of the tools adopted to support decision-making.

(Footnote 28 continued)

unstructured, and consequential … There is no cut and dried method for handling the problem because it hasn't arisen before, or because its precise nature and structure are elusive or complex, or because it is so important that it deserves a custom tailored treatment" Simon (1960, pp. 5–6).

[29]E.g., for the interpretation of rules to apply to each problem.

[30]E.g., those which are formalized in the manuals that rule planning and reporting.

Regarding the modeling approach, a distinction can be made between: transactional systems, "information feedback" support, and "learning support" models.

Transactional systems (Koutsoukis and Mitra 2003, p. 38) are models/tools supporting the accomplishment of structured tasks on which operational control is focused. Their perspective is data gathering. They aim to collect detailed and precise data for the fulfillment of current transactions (e.g. expenditure and revenue cycles, payroll systems, inventorying). Collecting and storing such data feeds the overall information system. In fact, both the models/tools supporting decision-making related to management and strategic control draw from such databases to elaborate information, to foster decision-making and systemic learning.

"Information feedback support" models, such as accounting and financial budgeting aim to provide decision-makers with specific information for the fulfillment of semi-structured tasks, related to management control, which is primarily based on a feedback approach (i.e. a *post facto* analysis of performance discrepancies between actual and budgeted results). The basic component of such models is the general ledger, which is the backbone of the accounting system, generating financial information that is portrayed through balance sheets, income and cash flow statements. It also supports product and process costing. Other components of "information feedback support" models are related to financial budgeting and reporting. Though such tools may not be fully based on accounting data, they are usually focused on a financial dimension of organizational performance. Examples can be related to spreadsheet modeling and cost accounting.

A third component of the information system, supporting unstructured (and partially also semi-structured) decision-making to implement a strategic control that is systemically connected to management control can be defined as "learning support" models. Such models aim to foster strategic organizational learning to deal with uncertainty and discontinuity. To this end, they enhance organization's capabilities to manage dynamic complexity and to corroborate the traditional feedback mechanisms on which management control is based, with proactive feedforward control loops (De Hass and Kleingeld 1999, pp. 242–245; Hofstede 1978, p. 451; Kloot 1997, p. 52; Otley 1999, p. 369; Schreyand Steinmann 1987).

Likewise "information feedback support" models, also "learning support" models draw 'coded' internal data from the organizational information system. However, in this context, a different approach is followed in searching for such data, and using it. Here, in order to generate information leading to strategic learning, the main concern is on the promptness, relevance, synthesis, dynamic pattern of behavior, and systemic consistency of such data, rather than on its (static) precision, or detail (Amigoni 1978). Furthermore, a wider use of external databases (e.g., related to market competition) and subjective/inductive data estimates, drawn from decision makers' experience and mental models (e.g., perception delays, behavioral functions, intangibles) characterizes "learning support" models.

Examples of such models can be related to a quite wide variety of tools that rely on artificial intelligence (Sauter 2010; Turban and Watson 1989), such as: Executive Information Systems (Singh et al. 2002), expert systems (Brandon 1990),

MODELING APPROACH   ## FOCUS

LEARNING SUPPORT MODELS

➢ System Dynamics
➢ Executive Information Systems
➢ Expert Systems
➢ Neural networks
➢ Fuzzy systems
➢ Decision Support Systems
➢ ....

**INFORMATION FEEDBACK
SUPPORT MODELS**

➢ Accounting
➢ Financial Budgeting

**TRANSACTIONAL
SYSTEMS**

➢ Expenditure cycle
➢ Revenue cycle
➢ Payroll
➢ ...

STRATEGIC
CONTROL

MANAGEMENT
CONTROL

OPERATION
CONTROL

UNSTRUCTURED
DECISIONS

SEMI-STRUCTURED
DECISIONS

STRUCTURED DECISIONS

INFORMATION SYSTEM

Learning

Decision-making

Information
provision

Data
gathering

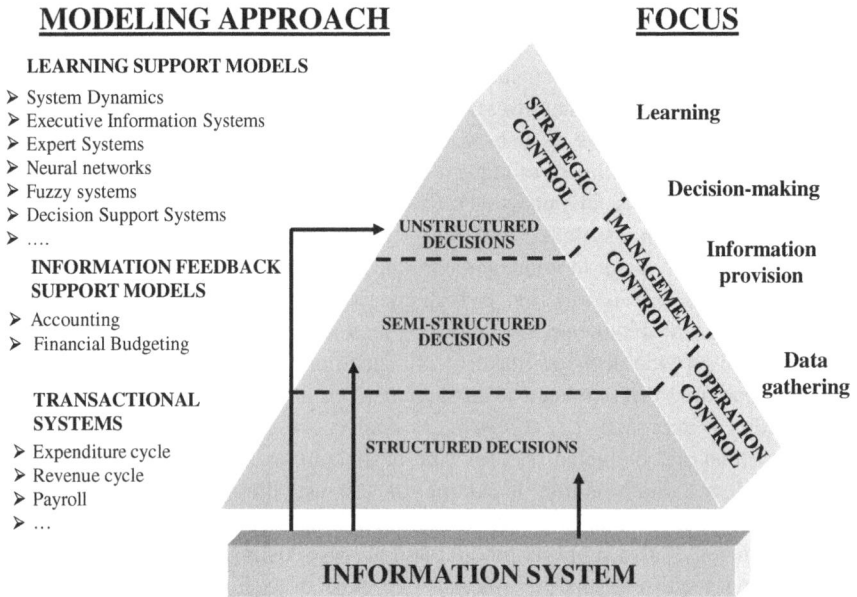

Fig. 1.14 Different modeling approaches supporting P&C systems and decision makers

neural networks, fuzzy systems (Bidgoli 1998), Decision Support Systems (Keen and Scott Morton 1978).

SD can be considered as a key component of such modeling approach. It can strongly enhance strategic planning and control. Also it can facilitate a learning and communication process by which strategic goals can be cascaded through the organization structure, and trade-offs over time and space can be framed. SD modeling can help controllers to design and operate consistent and learning-oriented P&C systems and to align strategic with management control.

Figure 1.14 frames the analysis developed so far in this section.

1.9 The Process of System Dynamics Modeling. Double Loop Learning in a Dynamic Performance Management Context to Pursue Sustainable Organizational Development

Figure 1.15 emphasizes the benefits that can be achieved by combining SD models with "information feedback support" models, to foster mental models' elicitation and improve organizational capabilities in assessing performance through a sustainable development perspective (dynamic performance management).

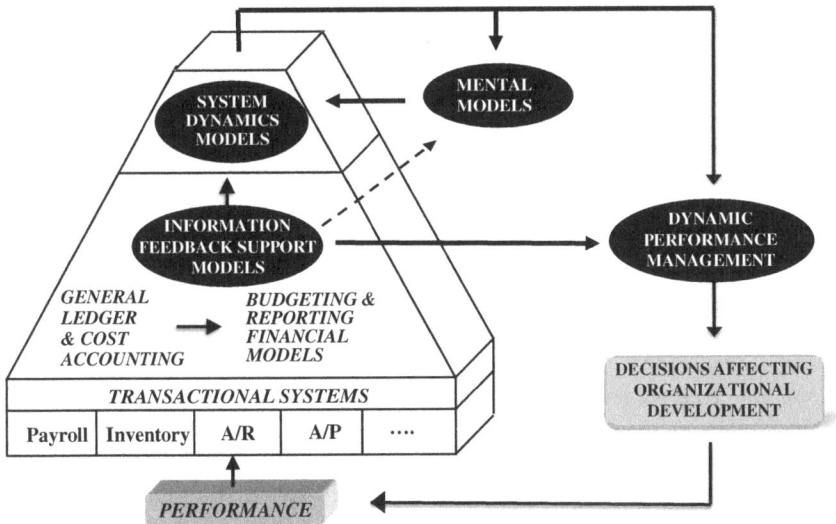

Fig. 1.15 Matching "information feedback support" models with SD models to foster mental models' elicitation and improve organizational capabilities to assess performance in a sustainable development perspective

Though this topic will be analyzed in more detail in the next chapters, it is possible here to remark how "information feedback support" models are mainly based on a financial perspective and on static performance measurement. Such perspective may not allow decision makers to properly handle dynamic complexity. To this end, a high selectivity and promptness is required to P&C systems. Not only capital investment decisions are to be considered as those having a potential impact on organizational sustainable growth, but also current decisions. In fact, though current management takes place on an on-going basis, not all current decisions have the same level of importance for the sustainable development of an organization. Detecting weak signals of strategic change hidden in current activities implies a level of complexity that is different from longer run decisions related to capital budgeting. Even though, in the first case, the structure of the system to manage can more easily be defined than in the second case, monitoring the strategic relevance of current events implies a major difficulty in detecting in advance weak signals of change. In fact, such signals are usually hidden in a wider range of daily occurrences in which decision makers can be fully involved.

Matching SD with "information feedback support" models may allow decision makers to frame the processes underlying accounting information. Drawing up plans only on the basis of single and static pieces of accounting information may perhaps allow organizations to design policies which are effective in the short run, but may also produce unintended negative effects on a longer time horizon, that seriously prejudice sustainable organizational development.

Understanding causal relationships laying behind organizational performance requires that double loop learning is fostered to enhance decision makers' aptitudes to detect inconsistencies in their mindsets and to pursue a common share view of the real world (Winch 1993).

Improving mental models by matching "information feedback support" financial models with SD models enhances dynamic performance management, through which decisions affecting development can be framed in an organizational sustainability perspective.

Figure 1.16 shows how decision makers' culture plays a central role in influencing mental models, decisions and actions (including the design of P&C and human capital development systems). In fact, the lens through which reality is observed depends on values, beliefs and other personal traits. For instance, a *technocratic* entrepreneur will be focused on the R&D and production areas, while an *autocratic* entrepreneur will not be inclined to perceive his/her competitors and workers as potential partners. At the same time, an entrepreneur who operates since the beginning of his/her activity in an *arena* based on price competition and low cost, will not be prone to think that among potential clients there may be some who could be more sensitive to delivery delay or product scope, rather than price.

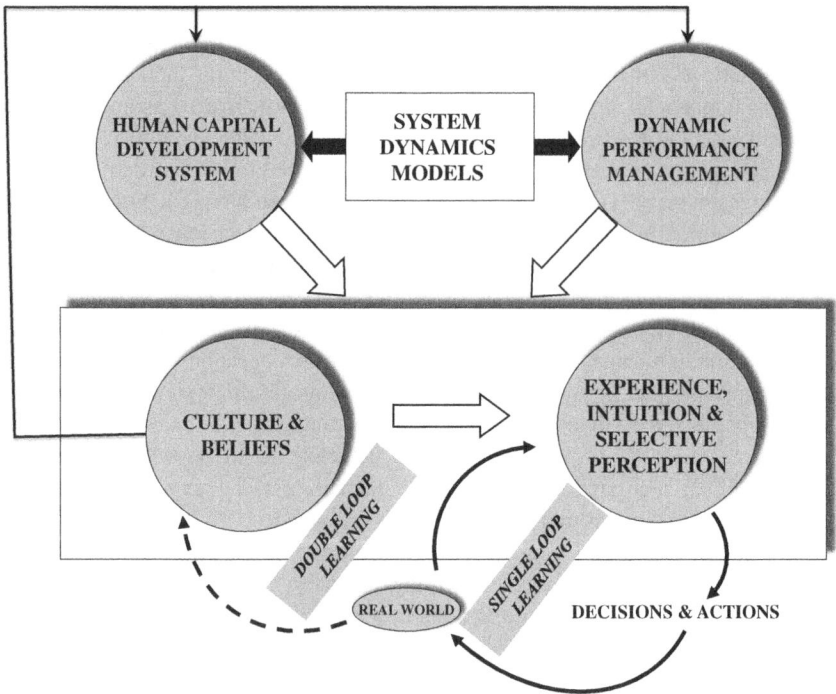

Fig. 1.16 Enhancing organizational learning through dynamic performance management and human capital development policies

Particularly when an organization operates in dynamic complex systems, decision makers may be blinded by their own mental models in promptly perceiving gradual changes occurring in the relevant system's structure and understanding the deep causes of organizational phenomena. Even though experience, intuition, intelligence and entrepreneurial skills may substantially support decision-making, cognitive limitations of human beings are often a primary cause of misperceptions concerning feedback loops and non-linear relationships between relevant variables.

On the other hand, P&C and education systems may play a significant role to affect organizational culture and decision makers' mental models. Business survival and sustainable development can be enhanced by providing decision makers with a relevant information set and strong learning support, to help them in eliciting and communicating how they map the relevant system, to assess their mental models' consistency.

Achieving a common shared view of the "real world" is not a symptom of conformism (i.e. forcing people to adopt a common vision); it is instead, a result of a learning process, which stems from the comparison and coherent combination of the variety of frames through which things are perceived by different players in an organizational context. Making mental models explicit and sharing them is not an end per se (Vennix 1996); it is, rather, a means through which people are helped to raise questions on relevant issues regarding trade-offs in time and space, related to sustainable organizational development. The main concern of learning in and about complex systems is not simply to find the 'right' solutions to problems, but instead to understand their deep causes (Sterman 1994).

Furthermore, learning should not to be conceived as a contingent or discrete activity (i.e., to be fulfilled through only ad hoc task forces), but instead as a continuous process. In fact, in a complex and dynamic context, 'freezing' such learning process in a bounded time horizon could not allow decision makers to gain an ability to affect organizational outcomes.

Improving organizational key-actors' mental models is the goal of a continuous learning process on which dynamic performance management should be focused. Such goal can be pursued through an activity requiring a continuous effort, fostered by a SD modeler/learning-facilitator, aimed to feed a circular process articulated along the following interconnected phases: (1) observation; (2) reflection, knowledge elicitation and communication; (3) diagnosing and sharing of a same picture of reality; (4) decision making and action (Kolb 1984).

The process of SD modeling can support the performance management cycle through the following phases: (1) mapping (framing the system); (2) planning; (3) implementing decisions/operations; (4) measuring/evaluating results, and undertaking corrective actions (Fig. 1.17).

SD modeling can support such cycle by fostering double loop learning. In fact, SD model building[31] starts with the identification of a dynamic pattern of behavior

---

[31]A comprehensive analysis of different approaches describing the SD modeling process can be fund in Martinez Moyano and Richardson (2013).

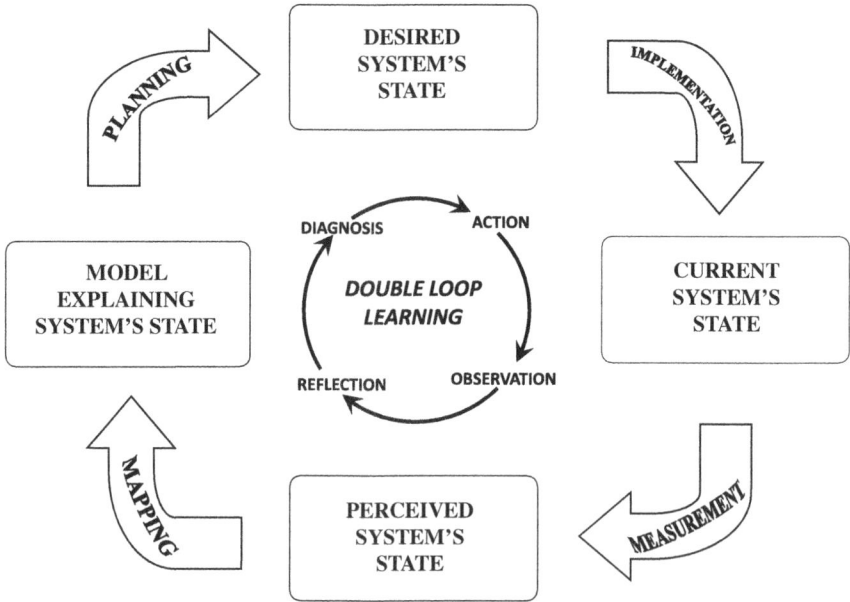

Fig. 1.17 Dynamic performance management and double loop learning

related to reported results (data collection/measurement). Such results are an effect of the current state of the organizational system. The goal of dynamic performance management is to: (1) perceive such state, (2) sketch—through system's mapping— a model that may explain the hidden feedback structure underlying detected behavior, (3) outline—through planning—a desired system's state, (4) implement the plan and undertake corrective action through feedback and feedforward control.

As previously said, the source of this data is not only from accounting or other organizational databases. It can be also related to decision makers' perceptions or even external datasets.

The identification of a dynamic problem behind detected reference behavior modes over time, allows decision makers (through the support of the SD modeler) to sketch a picture of the perceived system's state. Such picture initially frames the system through qualitative feedback analysis (i.e. influence diagrams are sketched and shared among decision makers). Hypotheses on systems structure and behavior are formalized. In a second step of SD modeling supporting planning, stock and flow quantitative analysis is developed. As previously remarked, depending on the scope of the analysis, as well as on the level of knowledge/SD culture and information an organization holds, this step might be limited to the development of an insight simulation model. Through model validation, and structure and behavior analysis during simulation, decision makers can review their own initial hypotheses. Based on this activity, dynamic plans can be sketched, and decisions can be made.

Double loop learning is enhanced through all the described process. The development, validation and use of simulation models supporting planning (with the related "reflection-diagnosis" stages), and the measurement of achieved results (with the related "implementation-measurement" stages) that will support further improvements to the original model are the cornerstones of a continuous learning process.

1.10 Applying System Dynamics Modeling to Performance Management: Different Contexts in Managing Organizational Sustainable Development and Restructuring Processes

The "structure and behavior" principle, we have analyzed in this chapter, can be helpful to outline possible contexts where SD modeling may successfully support performance management in different stages of organizations' life, to frame dynamic complexity.

Two main generic contexts can be distinguished on this regard:

1. Supporting decision makers to assess *current* performance levels, and to diagnose the current scenario patterns underlying the existing state of an organization, and
2. Supporting decision makers to assess *future* performance levels, where planning is expected to enhance the outline and implementation of strategies to manage sustainable organizational growth or restructuring policies to fix crises.

The two generic contexts are strongly related to each other. In fact, as previously remarked, diagnosing the current system's state is a fundamental step for policy making.

More specifically, regarding the first generic context, Fig. 1.18 frames four scenarios where SD modeling can be applied to support organizational performance management, i.e.:

(a) Growth;
(b) Restoring capacity;
(c) Crisis, and
(d) Divesting capacity.

The four scenarios emerge by matching the current strategic perspective of the organization with the polarity of the dominant feedback loops explaining the observed system's behavior. They can be both related to the overall organization system or to parts of it (e.g. strategic business areas).

Concerning the current strategic perspective of an organization, one may distinguish two opposite conditions, i.e.: Stability/Growth *versus* Downsizing.

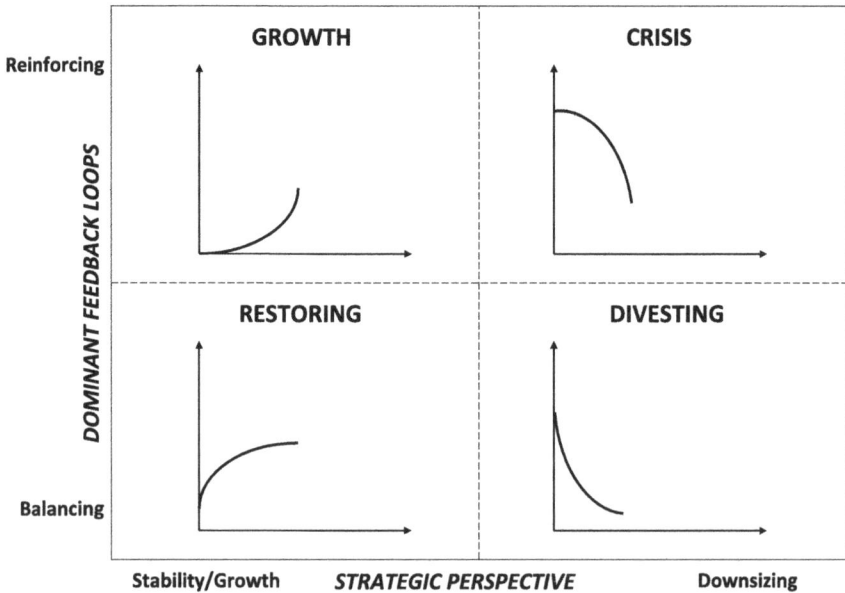

Fig. 1.18 Main scenarios associated to applying SD modeling to frame the current performance of an organization

If we take the first condition, those organizational systems whose current behavior is characterized by reinforcing loops underlying exponential dynamics in the observed results (e.g. sales, change in the customer base, cash flows, profits, change in business image) can be related to a "Growth" scenario.

A second scenario, implying a strategic perspective characterized by the search for stability, can be defined as "Restoring Capacity". This scenario underlies a dominance of balancing feedback loops aiming to adjust the lost capacity (e.g. staff, machinery, distribution network) to the desired levels.

If we consider the second condition (i.e., downsizing), those organizational systems whose current behavior is characterized by reinforcing loops underlying exponential dynamics in performance measures can be referred to a "Crisis" scenario. In this case, the current state of the organization system is affected by a progressive pattern of past negative results (e.g., in terms of loss of image, customers, staff, employee motivation, profitability, liquidity) that have been shrinking the stock of strategic resources.

The fourth scenario, related to a current downsizing perspective, can be defined as "Divesting Capacity". It implies a dominance of balancing feedback loops aiming to drain capacity (e.g. staff, machinery, distribution network).

Regarding the second generic context (i.e. supporting decision makers to assess future performance levels), Fig. 1.19 frames four specific scenarios where SD modeling can be applied:

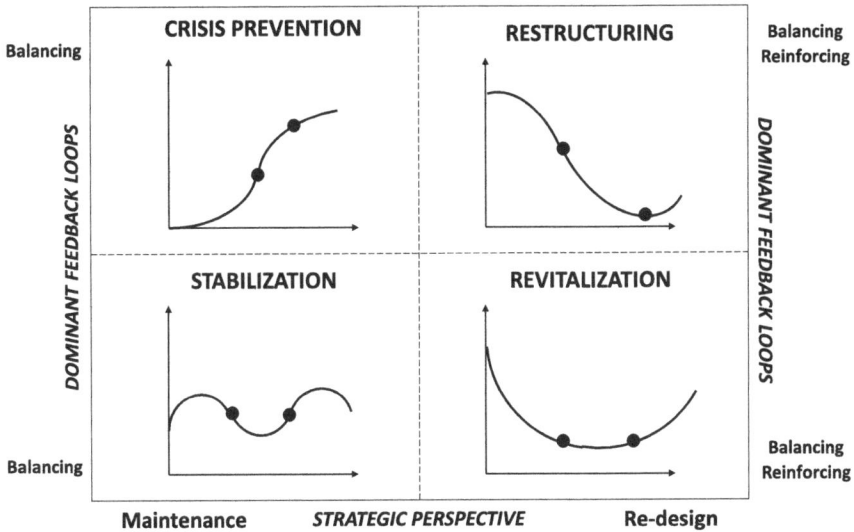

Fig. 1.19 Main scenarios associated to applying SD modeling to frame the future performance of an organization

(a) Crisis prevention;
(b) Stabilization;
(c) Restructuring, and
(d) Revitalization.

Each of the four scenarios can be considered as a logical extension of the previously commented scenarios, regarding the current system's state. They are associated with two possible strategic perspectives for the future, i.e.: Maintenance (i.e. preserving or consolidating current performance and the related system's structure) *versus* Re-design (i.e. changing the "logic model" behind current performance and the related system's structure).

If we take the first strategic perspective, a "Crisis prevention" scenario can be related to the use of SD modeling to support the continuous scanning of early crisis symptoms that could be generated by unsustainable growth rates. As Chap. 3 will illustrate, such symptoms are performance drivers that should be monitored in the short run, in order to prevent crises in the long run. Examples can be related to dominant (but still "silent", at least in terms of effects on the end-results) balancing loops associated with capacity bottlenecks or market saturation.

A second scenario can be defined as "Stabilization". It implies the pursuit of a non-growth condition, based on the same principles we previously commented about the "Restoring" scenario. "Stabilization" can be either intended to consolidate a competitive or financial position gained from past growth-oriented strategies, or to maintain the current performance levels that in the past have been made possible by adjusting capacity to the desired levels.

"Restructuring" is a third scenario. In this context, SD modeling may support policy makers in counteracting a current crisis condition. The dominant feedback loops related to the outline of restructuring strategies can be profiled according to two sequential stages, i.e. detecting and enhancing: (1) balancing loops that would allow the organization to restore conditions that have been lost due to crisis (e.g. financial resources, capacity), and (2) reinforcing loops that will allow the organization to pursue new growth, after that the dominance of adjusting balancing loops will have restored the lost dynamic equilibrium. Such reinforcing loops might foster, for instance, commercial or R&D growth, or the development of intellectual capital, or positive stakeholder relationships that should enable the firm to pursue further financial growth.

A fourth scenario is "Revitalization". This is related to a current "Divesting" condition, for instance concerning a specific strategic business area of a firm that needs to be re-designed in order to generate future growth. In this context, SD modeling can support decision makers to identify: (1) the draining balancing loops that should be enhanced in the short run, to accelerate the dissipation of those resources that are no more consistent with the new system structure (e.g. the business model) that should be re-designed, and (2) the reinforcing loops that should be enhanced in the long run, to foster new growth.

As shown in Fig. 1.19, for each of the four commented scenarios, SD modeling may help decision makers in identifying the *turning points* in system's behavior. Such *turning points* would require prompt and purposeful action, to enhance the dominance of feedback loops that will allow an organization to successfully design and implement strategy. In terms of strategic and management control, this means providing the P&C system with stronger feed-forward mechanisms that would enhance the ability of performance management to deal with the specific dynamic complexity embodied by each scenario.

For instance, "Crisis prevention" might imply the identification of diminishing returns from commercial efforts (in terms of sales order rates), due to market saturation.

In the "Stabilization" scenario, turning points might be identified into negative gaps between current (e.g. capacity, or product portfolio average age) and normal resource levels, which would allow the firm to preserve a stable performance condition.

"Restructuring" might imply the identification of turning points associated with the emerging effects on performance that could be first originated by the balancing loops and later by the reinforcing loops enhanced by adopted policies. For instance, a reduction in the financial losses could be generated in the short run by a lower "debts-to-investments" ratio caused by equity investments, leading to diminishing interest costs. Also, an increase in the customer base could be obtained in the long run through R&D policies, enabled by the new equity investments, leading to improvements in product performance.

A "Revitalization" scenario might imply the identification of turning points related to the draining effects in the short run generated by balancing loops on the resources that are no more relevant for the new designed business model (e.g. competencies, technologies, suppliers). It might also imply the setting of turning points related to the acquisition of new sources of competitive advantage that can be fostered in the long run by company policies aimed to enhance reinforcing loops (e.g. new technology or market communication investments).

References

Alford, J., & O' Flynn, J. (2009). Making sense of public value: Concepts, critiques and emergent meanings. *International Journal of Public Administration, 32*, 171–191.

Amigoni, F. (1978). Planning management control systems. *Journal of Business Finance and Accounting, 5*(3), 279–291.

Andersen, D. (1990). Analyzing who gains and who loses: The case of school finance reform in New York State. *System Dynamics Review, 6*, 21–43.

Ansoff, H. I. (1965). *Corporate strategy: An analytic approach to business policy for growth and expansion.* New York: McGraw-Hill.

Anthony, R. (1965). *Planning and control systems: A framework for analysis.* Boston: Harvard Business School Division of Research.

Anthony R., & Govindarajan, V. (2007). Management control systems. Chicago: Mc-Graw-Hill/Irwin.

Argyris, C. (1985). *Strategy.* Pitman, Boston: Change and Defensive Routines.

Argyris, C., & Schon, D. (1978). Organizational learning. A theory of action perspective. Reading, Mass: Addison Wesley.

Arthur, D., & Winch, G. W. (1999) Extending model validity concepts and measurements in system dynamics. In *Proceedings of the 1999 International System Dynamics Society Conference.* Wellington, NZ.

Barlas, Y., & Diker, V. (2000). A dynamic simulation game (UNIGAME) for strategic university management. *Simulation & Gaming, 31*, 3, 331–358.

Bernstein, D. (1994). Dynamics of the New York city court system. In *International System Dynamics Conference Proceedings* (pp. 29–41).

Bhattacharyya, S. (1971). Considerations in designing management control systems. *Economic and Political Weekly, 6*(9), 9–16.

Bianchi, C. (2002). Introducing SD modeling into planning and control systems to manage SMEs' growth: A learning-oriented perspective. *System Dynamics Review, 18*(3), 315–338.

Bianchi, C. (2010). Improving performance and fostering accountability in the public sector through system dynamic modeling: From and 'external' to an 'internal' perspective. *Systems Research and Behavioral Science, 27*, 361–384.

Bianchi, C., & Rivenbark, W. C. (2014). Performance management in local government: The application of system dynamics to promote data use. *International Journal of Public Administration, 37*, 13, 945–954.

Bianchi, C., & Williams D. (2015). Applying system dynamics modeling to foster a cause-and-effect perspective in dealing with behavioral distortions associated with a city's performance measurement programs. *Public Performance Management Review, 38*, 395–425.

Bianchi, C., & Winch, G. W. (2006). Unleashing growth potential in 'stunted' SMEs: Insights from simulation experiments. *International Journal of Entrepreneurship and Small Business, 3*, 1, 92–105.

Bianchi, C., Winch, G. W., & Tomaselli, S. (2008). Management simulation as an instrument to aid turning "stunted growth" round in family businesses. *Sinergie, 75*, 109–126.

Bidgoli, H. (1998). *Intelligent management support systems.* Westport: Quorum Books.

Blau, P. M., & Scott, W. (1962). Formal organizations. Chandler, San Francisco.

Boyle, R. (1999). The management of cross-cutting issues. Discussion paper no. 8, committee for public management research. Dublin: Institute of Public Administration.

Borgonovi, E. (2001). Il concetto di valore pubblico [*The Concept of Public Value*]. *Azienda Pubblica, 2*(3), 185–188.

Brandon, P. (1990). The development of an expert system for the strategic planning of construction projects. *Construction Management and Economics, 8*, 285–300.

Busco, C., & Riccaboni, A. (2010). Performance management systems and organizational culture. In P. Taticchi (Ed.), *Business performance measurement and management* (357–376).

Chenhall, R. (2003). Management control systems design within its organizational context: Findings from contingency-based research and directions for the future. *Accounting, organizations and society, 2–3*, February–April, 127–168.

Christensen, T., & Laegreid, P. (2007). *Transcending new public management*. Surrey: Ashgate.

Coda, V. (2010). *Entrepreneurial values and strategic management. Essays in management theory*. New York: Palgrave Mac Millan.

Cohen, M. D., March, J. G., & Olsen, J. P. (1972). A garbage can model of organizational choice. *Administrative Science Quarterly, 17*, 1, 1–25.

Coyle, R., & Alexander, M. (1997). Two approaches to qualitative modeling of a nation's drugs trade. *System Dynamics Review, 13*(3), 205–222.

Cyert, R. M., & March, J. G. (1963). *A Behavioral theory of the firm*. New Jersey, Prentice Hall.

Dagnino, G., & Quattrone, P. (2006). Comparing institutionalisms Gino Zappa and John R. Commons' accounts of "institution" as a groundwork for a constructivist view. *Journal of Management History, 12*, 1.

Das, S. (2011). *Management control systems: Principles and practices*. New Delhi: PHI Publishers.

Davidsen, P. (1991). *The structure-behavior graph. System dynamics group*. Cambridge, MA: MIT.

Davidsen, P., Sterman, J., & Richardson, G. (1990). A Petroleum Life Cycle Model for The United States with Endogenous Technology, Exploration, Recovery, and Demand. *System Dynamics* Review, 6, 1, p. 66–93.

De Geus, A. (1994). Modeling to predict or to learn? In J. D. Morecroft & J. D. Sterman (Eds.), *Modeling for learning organizations*. Portland: Productivity Press.

De Geus, A. (1997). *The Living Company*. Habits for Survival in a Turbulent Business Environment: Harvard Business School Press.

De Haas, M., & Kleingeld, A. (1999). Multilevel design of performance measurement systems: Enhancing strategic dialogue throughout the organization. *Management Accounting Research, 10*, 1, 233–261.

Dorner, D. (1980). On the difficulties people have in dealing with complexity. *Simulation and Games, 11*(1), 87–106.

Dorner, D. (1997). *The logic of failure*. Reading, MA: Addison-Wesley.

Dorner, D., & Schaub, H. (1994). Errors in planning and decision-making and the nature of human information processing. *Applied Psychology, 43*, (4), 433–453.

Dyner, I., & Larsen, E. (2001). From planning to strategy in the electricity industry. *Energy Policy, 29*, 13, 1145–1154.

Ferreira, A., & Otley, D. (2009). The design and use of performance management systems: An extended framework for analysis. *Management Accounting Research, 20*, 263–282.

Festinger, L. (1957). *A theory of cognitive dissonance*. Stanford, CA: Stanford University Press.

Fiddaman, T. (2007). Dynamics of climate policy. *System Dynamics Review, 23*(1), 21–34.

Flamholtz, E. (1983). Accounting. Budgeting and control systems in their organizational context: Theoretical and empirical perspectives. *Accounting, Organizations and Society, 8*(2–3), 153–169.

Flamholtz, E. (1986). Effective management control: A framework. *Applications and Implications, European Management Journal, 14*(6), 596–611.

Flamholtz, E. (1996). Effective organizational control: A framework, applications, and implications. *European Management Journal, 46*(6), 596–611.

Flamholtz, E., Das, T., & Tsui, A. (1985). Toward an integrative framework of organizational control. *Accounting, Organizations and Society, 10*, 1, 35–50.

Ford, A. (1999). *Modeling the environment*. Washington: Island Press.

Forrester, J. W. (1961). *Industrial dynamics*. Cambridge, Mass: MIT Press.

Forrester, J. W. (1968). Market growth as influenced by capital investment. *Industrial Management Review, 9*(2), 83–105.

Forrester, J. W. (1969). *Urban dynamics*. Waltham: Pegasus.

Forrester, J. W. (1970). *World dynamics*. Cambridge, MA: MIT Press.

Forrester, J. W. (1971). Counterintuitive behavior of social systems. *Technology Review, 73*(3), 52–68.

Forrester, J. W. (1985). The model versus a modeling "Process". *System Dynamics Review, 1*(1), 133–134.

Forrester, J. W. (1994). Policies, decisions and information sources for modeling. In Morecroft, J. D., & Sterman, J. D. (Eds.), *Modeling for learning organizations*. Portland: Productivity Press.

Ghaffarzadegan, N., Lyneis, J., & Richardson, G. P. (2011). How small system dynamics models can help the public policy process. *System Dynamics Review, 27*, (1), 22–44.

Goodman, M. (1989). *Study notes in system dynamics*. Boston, MA: Productivity Press.

Grossler, A. (2004). A content and process view on bounded rationality in system dynamics. *Systems Research & Behavioral Science, 21*(4), 319–330.

Grossler, A. (2008). System dynamics modelling as an inductive and deductive endeavour. *Systems Research & Behavioral Science, 25*(4), 467–470.

Grynkewich, A., & Reifel, C. (2006). Modeling jihad: A system dynamics model of the salafist group for preaching and combat financial subsystem. *Strategic Insights, 5*(8), 1–26.

Jaen, S., & Dyner, I. (2008). Criminal cycles in the illegal drug industry: A system dynamics approach applied to colombia. *Winter Simulation Conference*, WSC (pp. 1429–1436).

Hamel, G., & Prahalad, C. (1994). Competing for the future. Boston: Harvard Business School Press.

Hofstede, G. (1978). The poverty of management control philosophy. *The Academy of Management Review, 3*(3), 450–461.

Hofstede, G. (1980). Management control of public and not-for-profit activities, working paper. In *International Institute for Applied Systems Analysis*. Laxenburg, Austria (downloadable from: http://webarchive.iiasa.ac.at/Admin/PUB/Documents/WP-80-052.pdf)

Homer, J. (1993). A system dynamics model of national cocaine prevalence. *System Dynamics Review, 9*, 49–78.

Honggang, X. (2003). Managing side effects of cultural tourism development: The case of Zhouzhuang. *Systems Analysis Modeling Simulation, 43*(2), 175–188.

Kaplan, R., & Norton, D. (2001). The strategy-focused organization. Boston: Harvard Business School Press.

Kaplan, R., & Norton, D. (2008). *The execution premium: Linking strategy to operations for competitive advantage*. Boston: Harvard Business School Press.

Keen, P., & Scott Morton, M. (1978). *Decision support systems. An organizational perspective*. Reading, Mass: Addison Wesley.

Kloot, L. (1997). Organizational learning and management control systems: Responding to environmental change. *Management Accounting Research, 8*, 47–73.

Kolb, D. (1984). *Experiential learning*. Englewood Cliffs: Prentice Hall.

Kopainsky, B., Pedercini, M., Davidsen, P., & Alessi, S. (2009). A blend of planning and learning: Simplifying a simulation model of national development. *Simulation and Gaming*, 641–662.

Koutsoukis, N. S., & Mitra, G. (2003). *Decision modeling and information systems*. Kluwer, Norwell, Mass.

Lane, D. (1994). Modeling as learning: A consultancy methodology for enhancing learning in management teams. In J. D. Morecroft & J. D. Sterman (Eds.), *Modeling for learning organizations*. Portland: Productivity Press.

Lane, D. (2001). Rerum cognoscere causas: Part 2—opportunities generated by the agency/structure debate and suggestions for clarifying the social theoretic position of system dynamics. *System Dynamics Review, 17*(4), 293–309.

Lane, D., & Huseman, E. (2008). System dynamics mapping of acute patient flows. *Journal of the Operational Research Society, 59*, 213–224.

Lindblom, C. (1959). The science of "muddling through". *Public Administration Review, 19*(2), 79–88.

Lorange, P., & Scott Morton, M. S. (1974). A framework for management control systems. *Sloan Management Review, 16*, 1, 41–56.

Lorange, P., Scott Morton, M. F., & Ghoshal, F. (1986). *Strategic control*. St. Paul: West Publishing.

Maciariello, J. A. (1984). *Management control systems*. Englewood Cliffs: Prentice Hall.

Martinez Fernandez, J., & Esteve Selma, M. (2004). The dynamics of water scarcity on irrigated landscapes: Mazzarro'n and Aguillas in Southeastern Spain. *System Dynamics Review, 20*, 2, 117–137.

Martinez Moyano, I., & Richardson, G. (2013). Best practices in system dynamics modeling, *System Dynamics Review, 29*, 2, 102–123.

Matheson, A., Scanlan, G., & Tanner, R. (1997). Strategic management in government: Extending the reform model in New Zealand. In *Benchmarking, evaluation and strategic management in the public sector*. Available at http://www.oecd.org/newzealand/1902913.pdf

Meadows, D., Behrens, W., Meadows, D., Naill, R., Randers, J., & Zahn, E. (1974). *The dynamics of growth in a finite world*. Cambridge, MA: Wright-Allen Press.

Meadows, D., Fiddaman, T., & Shannon, D. (2001). Fish banks, Ltd. A micro-computer assisted group simulation that teaches principles of sustainable management of renewable natural resources. *Laboratory for Interactive Learning*. Duhram: University of New Hampshire.

Meadows, D., Meadows, D., & Randers, J. (1992). *Beyond the limits: Confronting global collapse; envisioning a sustainable future. Post Mills*. Vermont: Chelsea Green.

Mintzberg, H., & Waters, J. A. (1985). Of strategies, deliberate and emergent. *Strategic Management Journal, 6*, 3, 257–272).

Moore, M. (1995). *Creating public value. Strategic Management in Government*. Cambridge, Mass: Harvard University Press.

Moynihan, D. P. (2008). *The dynamics of performance management*. Washington DC: Georgetown University Press.

Moynihan, D. P., & Landuyt, N. (2009). How do public organizations learn? Bridging cultural and structural perspectives. *Public Administration Review, 69*, 6, 1097–1105.

Morecroft, J. (1985). Rationality in the analysis of behavioral simulation models. *Management Science, 31*(7), 900–916.

Morecroft, J. (1988). System dynamics and microworlds for policymakers. *European Journal of Operational Research, 35*, 301–320.

Morecroft, J. (1998). Resource system management and dynamic complexity. In *Fourth International Conference on Competence-Based Management*. Oslo: Norwegian School of Management.

Morecroft, J. (2002). Resource management under dynamic complexity. In J. Morecroft, W. Sanchez & R. Aimé (Eds.), *Systems perspectives on resources, capabilities, and management processes*. Bingly: Emerald.

Moxnes, E. (2005). Policy sensitivity analysis: Simple versus complex fishery models. *System Dynamics Review, 21*(2), 123–145.

Moxnes, E., & Saysel, A. K. (2009). Misperceptions of global climate change: Information policies. *Climatic Change, 93*, 1–2, 15–37.

Mussari, R., & Ruggiero, P. (2010). Public managers' performance evaluations systems and public value creation: Behavioral and economic aspects. *International Journal of Public Administration, 33*, 541–548.

Nilsson, F., Olve, N. G., & Parment, A. (2011). *Controlling for competitiveness. Strategy formulation and implementation through management control*. Malmo: Copenhagen Business School Press.

Nklit, H. (2000). The balanced scorecard. A critical analysis of some of its assumptions. *Management Accounting Research, 11*(1), 65.

Osborne, D., & Gaebler, T. (1993). Reinventing government. Toronto: Penguin Books.

Otley, D. (1994). Management control in contemporary organizations: Towards a wider framework. *Management Accounting Research, 5*, 289–299.

Otley, D. (1999). Performance management: A framework for management control systems research. *Management Accounting Research, 10*(10), 363–382.

Otley, D. (2001). Extending the boundaries of management accounting research: Developing systems for performance management. *British Accounting Review, 33*, 241–243.

Ouchi, W. (1979). A conceptual framework for the design of organizational control mechanisms. *Management Science, 25*(9), 833–848.

Quinn, J. B. (1978). *Strategic change: "logical incrementalism"* (pp. 7–19). Fall: Sloan Management Review.

Richardson, G. P. (1991). *Feedback thought in social science and systems theory*. Waltham, MA: Pegasus.

Richardson, G. P. (2011). Reflections on the foundations of system dynamics. *System Dynamics Review, 23*(3), 219–243.

Richardson, G. P., & Lamitie, R. (1989). Improving connecticut school aid: A case-study with model-based policy analysis. *Journal of Education Finance, 15*, 169–188.

Richardson, G. P., & Pugh, A. L. (1981). *Introduction to system dynamics modeling*. Portland, Oregon: Productivity Press.

Richardson, G. P., Andersen, D., & Luna-Reyes, L. F. (2004). Joining minds: Group modeling to link people, process, analysis, and policy design. *Twenty-Sixth Annual APPAM Research Conference*. Atlanta, Oct 28–30.

Saeed, K. (1996). Sustainable development: Old conundrums, new discords. *System Dynamics Review, 12*(1), 59–80.

Sauter, V. (2010). *Decision support systems for business intelligence*. Hoboken, New Jersey: Wiley.

Schrey G., & Steinmann, H. (1987). Strategic control: A new perspective. *The Academy of Management Review, 12*, 1, 91–103.

Schwaninger, M., & Groesser, S. (2008). System dynamics as model based theory building. *Systems Research and Behavioral Science, 25*, 4, 447–465.

Shrage, M. (1991). Spreadsheet: Bulking up on data. In *Los Angeles times*. 11th April.

Senge, P. (1990). *The fifth discipline: The art and practice of the learning organisation*. New York: Doubleday.

Simon, H. (1945). *Administrative behavior*. New York: The Free Press.

Simon, H. (1957). Rationality and decision making. In *Models of man: Social and rational-mathematical essays on rational human behavior in a social setting (collected papers)*. Oxford: Wiley.

Simon, H. (1960). *The new science of management decision*. New York: Harper & Row.

Simon, H. (1979). Rational decision making in business organizations. *American Economic Review, 69*, 4.

Simons, R. (2000). *Performance measurement & control systems for implementing strategy*. Upper Saddle River, New Jersey: Prentice Hall.

Singh, S., Watson, H., & Watson, R. (2002). EIS support for the strategic management process. *Decision Support Systems, 33*, 1, 71–85.

Smith, P. (1995). On the unintended consequences of publishing performance data in the public sector. *International Journal of Public Administration, 18*, 2/3, 277–310.

Stephens, C., Graham, A., & Lyneis, J. (2005). System dynamics modeling in the legal arena: Meeting the challenges of expert witness admissibility. *System Dynamics Review, 21*, 2, 95–172.

Sterman, J. (1994). Learning in and about complex systems. *System Dynamics Review, 10*(2–3), 291–330.

Sterman, J. (2000). *Business dynamics. Systems Thinking and modeling for a complex world.* Boston: Irwin/McGraw Hill.

Sterman, J. (2002a). All models are wrong: Reflections on becoming a systems scientist. *System Dynamics Review, 18*(4), 501–531.

Sterman, J. (2002b). *System dynamics: Systems thinking and modeling for a complex world, MIT working paper series*, ESD-WP-2003–01.13.

Sterman, J. (2002c). The global citizen: Celebrating the life of Dana Meadows. *System Dynamics Review, 18*, 2, 101–310.

Sterman, J., & Richardson, G. (1985). An experiment to evaluate methods for estimating fossil fuel resources. *Journal of Forecasting, 4*, 197–226.

Thompson, J., Millstone, E., & Scoones, I., et al. (2007). *Agrifood system dynamics: Pathways to sustainability in an era of uncertainty, STEPS working paper 4.* Brighton: STEPS Centre.

Turban, E., & Watson, H. (1989). Integrating expert systems, executive information systems, and decision support systems. *DSS Transactions* (pp. 74–82). Pakistan: Institute of Management Sciences.

Vakkuri, J., & Meklin, P. (2006). Ambiguity in performance measurement: A theoretical approach to organisational uses of performance measurement. *Financial Accountability & Management, 22*, 3, pp. 235—250.

Van der Heijden, K. (2004). *Scenarios: The art of strategic conversation.* Chichester: Wiley.

Van Dooren, W., Halligan, J., & Bouckaert, G. (2010). *Performance management in the public sector.* New York: Routledge.

Vennix, J. (1996). *Group model building.* Chichester: Wiley.

Vennix, J., & Gubbels, J. (1994). Knowledge elicitation in conceptual model building: A case-study in modeling a regional dutch health care system. In J. Morecroft & J. Sterman (Eds.), *Modeling for learning* (pp. 121–145). Productivity Press: Portland, Oregon.

Warren, K. (2000). The softer side of strategy dynamics. *Business Strategy Review, 11*(1), 45–58.

Warren, K. (2008). *Strategic management dynamics.* New York: Wiley.

Winch, G. (1993). Consensus building in the planning process: Benefits from a "hard" modeling approach. *System Dynamics Review, 9*(3), 287–300.

Winch, G., & Joyce, P. (2006). Exploring the dynamics of building, and losing, consumer trust in B2C eBusiness. *International Journal of Retail & Distribution Management, 34*, 7, 541–555.

Wolstenholme, E. (1999a). A patient flow perspective of U.K. health services: Exploring the case for new 'intermediate care' initiatives. *System Dynamics Review, 15*(3), 253–271.

Wolstenholme, E. (1999b). Qualitative vs. quantitative modelling: The evolving balance. *Journal of the Operational Research Society, 50*(4), 422–428.

Zagonel, A. (2002). Model conceptualization in group model building: A review of the literature exploring the tension between representing reality and negotiating a social order. *International Conference of the System Dynamics Society Proceedings.* Palermo, Italy, 28 July–1 August. http://www.systemdynamics.org/conferences/2002/proceed/papers/Zagonel1.pdf

Zappa, G. (1927). *Tendenze nuove negli studi di Ragioneria, (New Tendencies in Accounting Studies).* Milan: Istituto Editoriale Scientifico.

Zappa, G. (1937). *Il reddito di impresa – Scritture doppie, conti e bilanci di aziende commerciali [The Income of the Firm – Double Entry Bookkeeping, Accounts and Reports of Commercial Organizations].* Milan: Giuffrè.

Chapter 2
The Need of a Dynamic Performance Management Approach to Foster Sustainable Organizational Development

2.1 Organizational Growth, Strategy and Performance

The concept of organizational growth concerns the strategic domain of management. It underlies the aptitude of an organization to attain a set of results leading to its long-term success and continuity.

Growth, strategy, and performance are strictly related concepts. Strategic decisions concern the constitution, improvement, or change of a set of structures, e.g. involving organizational, production, distribution, and cultural assets (Flamholtz 1996; Flamholtz and Hua 2002; Langfield-Smith 1997; Lorange and Vancil 1976; Mintzberg and Westley 1992; Schreyögg and Steinmann 1987; Wernerfelt 1984). Such decisions affect the relationships between an organization and its environment, to change performance (Henri 2006; Kloot 1997; Munro and Wheeler 1980; Simons 2000).

The strategic decisions' common denominator can be referred to the search of performance targets on a set of measures portraying a balanced and sustainable organization development (Fig. 2.1).

Organizational growth can be, first, considered as a qualitative—rather than purely quantitative phenomenon. In these terms, growth implies *development*, i.e. a learning process, enhancing synergies with stakeholders (Ackoff 1986; Coda 2010; Sorci 2007).

Organizational growth also can be studied under a quantitative (or dimensional) perspective. This can be framed under both a structural and operational viewpoint. Under the first viewpoint, growth is measured in terms of investment stocks, available in a given time. Under the second viewpoint, growth is measured in terms of flows—e.g. sales volumes or revenues, personnel turnover rate, change in machinery capacity or R&D investments. Such a different perspective of growth gauges the aptitude of an organization to increase its structural endowment of resources, over time (Fig. 2.2).

© Springer International Publishing Switzerland 2016
C. Bianchi, *Dynamic Performance Management*, System Dynamics
for Performance Management 1, DOI 10.1007/978-3-319-31845-5_2

Fig. 2.1 Organizational performance, strategy, and sustainability

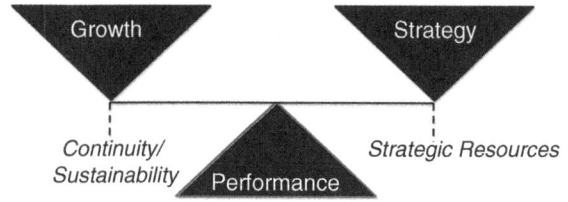

Fig. 2.2 Structural and operational viewpoints of organizational growth

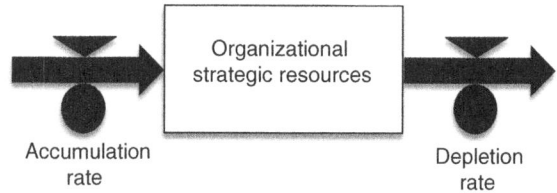

Though an organization may be affected by a lack of dimensional growth over a long time span, its survival and lifelong existence cannot disregard a continuous search for qualitative growth, i.e. development (Greiner 1972). Particularly in times of discontinuity, pursuing a hypothetical stable condition is a symptom of decline.

Every organization needs learning, which is—in turn—a pre-requisite for development and growth. Managing sustainable organizational development underlies an aptitude to match short with long-term, to combine efficiency with effectiveness (Coda 2010).

This chapter prepares the field for the analysis of a conceptual and methodological framework to support policy makers in framing and assessing performance within the perspective of sustainability. It emphasizes the need of a SD approach to enhance "intelligent" P&C systems so as to both manage performance and enhance sustainable development.

2.2   Three Perspectives of Sustainable Organizational Development

An organization's growth rate is balanced if it crosses different perspectives (Fig. 2.3). With respect to the organization itself, growth can be both assessed under an internal and an external profile.

Under the *internal profile*, balanced growth emerges from the search for consistency between different subsystems, sectors, and departmental/functional areas of an organization, or of a system of organizations (e.g. an industrial network or a system of public sector institutions). Therefore, an unbalanced growth rate could be associated with either a size increase or improvement in operations in one area of engagement to the detriment of another. For example, the performance of a strategic

Fig. 2.3 Framing
organizational growth: the
search for consistency
between the internal, external,
and time perspectives

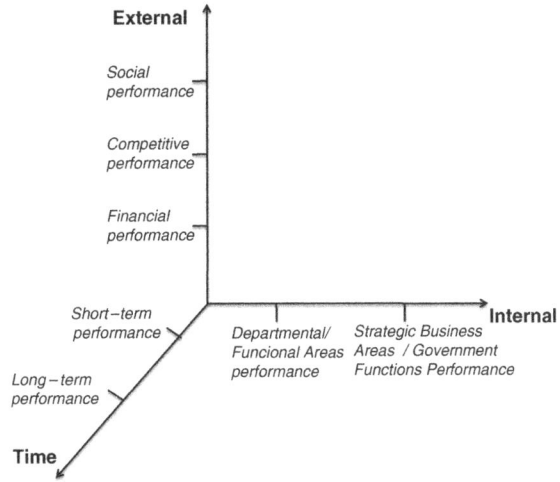

business area (SBA) in a company could be improved by diverting the resources invested in another SBA in the same company. Likewise, unbalanced growth in the public sector may imply a too intensive effort towards investments in an industry (e.g. chemical) to the detriment of another (e.g. tourism) in the same area.

Under the *external profile*, a balanced growth should be associated with performance rates crossing the three most relevant 'dimensions' of organizational success, i.e.: (1) financial; (2) competitive, and (3) social (Coda 2010). Such dimensions outline the physiological goals of an organization. The first dimension relates to the financial equilibrium and profitability, or at least to the balance between cash in-and-outflows in public and nonprofit organizations. The second one relates to the capability of an organization to satisfy its customers' needs with its products or services at a reasonable price, and therefore to generate value to the users' benefit. The third dimension expresses an organization's capability of meeting the expectations of its different stakeholders, e.g., workers, funders, and community.

Another perspective in which to assess sustainable growth is *time*. As discussed in the previous chapter, an improvement in short-term performance should not be obtained to the prejudice of long-term results. For instance, the recovery of company losses by indiscriminate investment-reduction policies, i.e., related to discretionary investments such as advertising or R&D, can adversely impact company profitability in the long run. Linking and balancing the short with the long term in planning and decision-making implies the need to adopt a strategic view of management. A strategic perspective is strictly related not only to classical long-term strategic planning "capacity decisions" but also to an analysis of the impact of current and often inertial decisions on the change in both organizational structures and external environmental conditions (Hamel and Prahalad 1994, p. XI).

Framing performance inclusively under financial, competitive, and social dimensions provides a reliable measure of organizational growth rate and sustainability. Sustainable growth underlies a performance rate that is consistent with all three perspectives, i.e., short versus long term, a given business area versus another, and the results in financial versus competitive versus social terms.

Leveraging on P&C systems to manage growth under the three said perspectives implies that the following questions are raised:

- Is the P&C system able to support the coordination of different levels of analysis, decision-making and responsibility? Is performance management carried out on only a bounded viewpoint or is there also a concern on whether (and how) different viewpoints can be made coherent with each other? Are strategic goals consistently cascaded to the lower levels of the organization? Is there a strategic dialog (both vertically and horizontally) among different organization levels?
- To what extent are decision makers made accountable on the outcomes produced by their own actions?
- To what extent are back-office units aware of their contribution to the wider organizational performance and to delivered product/service levels?
- Is the P&C system able to broaden the scope of analysis and evaluation from financial variables to the value generated by the organization into the wider socio-economic system?
- How is discontinuity perceived and managed? Does the P&C system enable decision makers to promptly perceive weak signs of strategic change, and to deal with them?
- Is the organization able to match in its current decision-making processes the need to keep strategic position with that of detecting and creating new growth opportunities?

There is the risk that P&C systems are designed and used according to a mechanistic, bureaucratic, and static perspective. Such risk may determine an *illusion of control*, rather than an enhanced capability of organizational decision makers to manage sustainable development, to promptly detect symptoms of crisis, to look for the causes of financial results, to set sustainable restructuring policies, to search for consistency in different subsystems, sectors, departments or governmental functions.

This issue has specific connotations in the public sector, where many reforms are still struggling to tackle consolidated cultures and practices, which are mostly focused on only the accomplishment of tasks and compliance to regulatory frameworks, legal prescriptions and procedures (e.g.: data gathering and processing, reporting results, calculating variances, respecting deadlines). Even an excessive concern on single technical tools (e.g.: information systems, accounting, or statistical computations) runs the risk of confusing means with ends, and misplacing focus on the design and use of consistent and "intelligent" P&C systems.

Today, because of resource scarcity and the proliferation of citizens' needs and expectations towards the public sector, P&C systems should be also focused on the search for continuous process improvement, and the measurement/management of outputs and outcomes. The need to outline strong political directives leading to viable performance plans requires that P&C analysts facilitate the planning process. They should play an active role inside political cabinets in fostering a strategic dialogue with administrative levels and in identifying—with the collaboration of departmental managers and the support of management control units—the operational objectives, as well as performance measures on which administrators should be made accountable, to attain the strategic goals.

Unfortunately, lack of P&C staff and performance management skills are often a major cause of blurred and of disconnect/unclear operational objectives formalized in the performance plans. It is not uncommon for operational objectives to be defined merely as activity descriptions, rather than measurable targets (Bianchi and Rivenbark 2012; Bianchi and Xavier 2014).

2.3 Framing Organizational Growth Sustainability: The Institutional and Interinstitutional Levels

Although the origin of sustainability studies can be seen in the biological sciences, more than to the managerial ones, a growing interest in the application of sustainability principles to the management of organizations is evident now.

The literature on balanced scorecards and corporate social responsibility provides empirical evidence of this phenomenon (Kaplan and Norton 1996; Werther and Chandler 2006), which is due to the rising dynamic complexity of the systems in which decision makers now operate. It also can be associated with the scarcity of available resources, both inside and outside a single organization.

Also, the topic of evaluating organizational performance within a sustainability perspective is grabbing more and more attention in the field of P&C studies and applications (Dyson 2000; Radermacher 1999; Riccaboni and Leone 2010). One may envisage two related levels for managing organizational performance under the perspective of sustainability, i.e., an *institutional* and an *inter-institutional* level.

At the institutional level, performance is assessed primarily in relation to the effects produced by decision makers on their own institution. At the inter-institutional level, performance is assessed in relation to the effects produced by decision makers on the wider system, e.g., either a local area or the industry to which they belong (Bianchi 2010).

Assessing organizational performance on an institutional level maintains a traditional viewpoint when growth sustainability is evaluated for a business. With respect to an enterprise, performance is associated primarily with company results, e.g., sales orders, revenues, income, and cash flows. However, today, due to increasing dynamic complexity in the competitive and social systems where businesses operate, firms perceive a growing need to assess their performance also at an inter-institutional level, e.g., when an enterprise takes a leading role in undertaking vertical or horizontal strategic relationships with other firms located in the value chain of its competitive system.[1]

Assessing business performance at an institutional level is a first step toward assessing performance at an inter-institutional level. In fact, a business that is able to combine the generation of profits with the creation of new employment, or of creating new industrial knowledge while increasing product quality at a reasonable price, is likely to contribute positively to the generation of value for the wider system. Such value will be measured in terms of tax contributions, increasing employment, shared knowledge with business partners, etc. This wider-system value will provide the conditions for the generation of new value to the benefit of each institution, and hence will generate new growth on an institutional level. So business growth can be considered as sustainable in the long run only if the firm generates value to the benefit of its local area or industry.

The relevant system's boundaries for such analysis are much broader than those associated with an institutional perspective. In fact, other public and private institutions are involved in such a system.

In an inter-institutional system perspective, assessing performance sustainability requires not only a focus on the single organization's results, but also on how such results contribute to the wider system's performance, a factor that will affect the organization in the long run. Inside such a wider system, each organization can build or share with others a given endowment of strategic resources (e.g., infrastructures, human capital, capacity, image, and environment). Both the aggregate performance of a local area or industry and the specific performance of each organization inside it are significantly affected by the accumulation and depletion processes of social capital[2] and other strategic resources, e.g., infrastructures and image. For instance, an opportunistic business behavior oriented to maximizing profits in the short run (e.g., without taking into account environmental pollution or human capital development issues), will contribute to depleting the quality of the local area's social capital and other strategic resources. In the long run, this will reduce the attractiveness and productivity of the region itself. A lower attractiveness could be measured, for instance, in terms of a negative market labor-turnover rate (resulting from the loss of population); a lower productivity could be measured in

---

[1]"Accountability may have to be interpreted as the development of mutual accountabilities between different organizational participants (*and indeed between different organizations*) rather than as solely a hierarchical process" (Otley 1994, p. 297. Italics added).

[2]Social capital refers to the connections among individuals and organizations, and to the norms of reciprocity and trustworthiness arising from them (Putnam 2000). Social capital is not just the sum of the institutions in a society; it is rather the glue that holds them together.

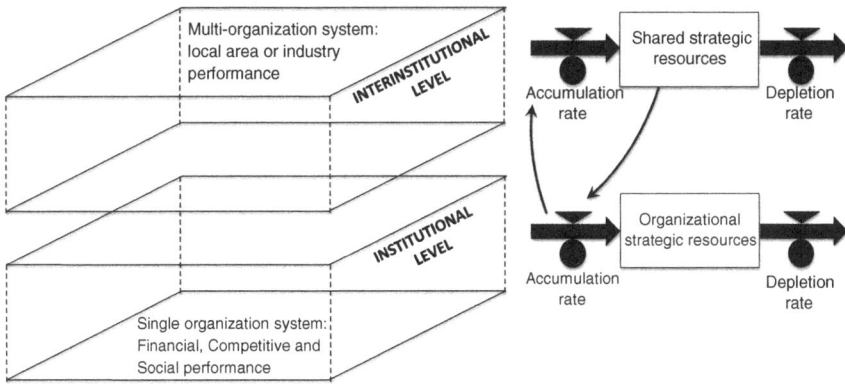

Fig. 2.4 The institutional and inter-institutional levels for analyzing organizational growth sustainability

terms of yield reduction in the exploited local resources (e.g., labor, raw materials, suppliers, and funders), and a drop in the level of synergy/collaboration between different actors in the system. A reduction in the local area's performance will also determine—sooner or later—a reduction in the performance of the opportunistic firm.

Figure 2.4 shows that strategic resources can be modeled as stocks of available tangible or intangible assets at a given time. Their dynamics depend on the value of corresponding in-and-outflows over time. Such flows are modeled as "valves" which decision-makers can regulate through their policies, to influence the dynamics of each strategic asset and therefore—through them—organizational performance at both the institutional and inter-institutional levels (Morecroft 2007; Warren 2008).

Managing strategic resources to affect performance is a dynamic and complex task. In fact, intangible resources (e.g., organizational climate, trust, knowledge, and image) are difficult to identify and measure. Furthermore, processes of accumulation and drain affecting the dynamics of strategic resources are inertial, since delays underlying them are difficult for decision-makers to perceive, and also because effects generated by actions taken (or not taken) in a recent or remote past are intertwined with each other, and single causes cannot be easily matched to related effects.

A tipping point in managing strategic resources to affect organizational performance is associated with the capability of policy-makers to (a) identify those strategic resources that most determine success in the environment (i.e., competitive and social systems) where an organization or different organizations operate, (b) insure that the endowment of such resources is satisfactory over time, and (c) keep a proper balance between the different relevant strategic resources.

2.4   Framing Sustainable Growth Within the Inter-institutional Level: Implications for Public Management

Framing organizational growth sustainability at an inter-institutional level should be a fundamental viewpoint to assess policy outcomes in public sector organizations. Particularly, implementing performance management in local government requires that an outcome view be adopted. This may allow local governments: (1) to assess the sustainability of their strategic plans and budgets, (2) to evaluate service delivery, and (3) to explore possible collaborative partnerships between different institutions in the same region for generating overall public value.

The path toward an outcome-based performance management in the public sector, however, is still difficult both from a theoretical and practical perspective, given the amount of effort involved in designing and operating performance management systems that may frame the public sector's specific complexity (Rainey and Han Chun 2005).

Concerning this issue, while performance management provides a wide area on which both research and practice have been working with specific reference to the private sector since a long time ago, it seems that many experiences matured over the years from success and failure in this field cannot be easily transposed to the public sector (Talbot 2005). In fact, the public sector is a complex and dynamic system, which is characterized by specific features. It is complex since several institutions (whose roles and competences cover different inter-related domains) affect performance. Complexity also stands into the constraints imposed to the public sector decision makers by the existing legal framework. Their decisions must always comply with such framework, although diverging from them could imply the achievement of better performance levels—e.g. in terms of efficiency and effectiveness.

The public sector is also a dynamic system, since the effects produced on performance by decisions made by the several (public and private) actors having a stake on the system itself, can be often observed after long delays. Such delays are due to the time it generally takes public sector decisions to generate their own outcomes on the community. They also depend on the huge net of feedback relationships between different subsystems (for instance, infrastructures may affect commerce or tourism, and in turn commerce or tourism can affect banking and—through this last subsystem—infrastructure funding, in a given Urban Metropolitan area).

Public sector performance has a major impact on the quality of life and may constitute either an acceleration factor or a constraint for the growth of the socio-economic sectors profiling a local area. A higher accountability of the public sector, and capability to deliver better services and rules to the private sector and the community, may generate economic and social value, in the system (Moore 1995). Such value corresponds to an increase in tangible and intangible strategic resources (e.g., infrastructures, funding, local area image, skilled workforce) that are available to the private sector. An improvement in such resources may result into a multiplier of the private sector performance, i.e. can determine suitable conditions to deliver products and services that can generate new value. Part of this value may, in turn, feed back to the public sector again, not only in terms of taxes and other financial contributions but also in terms of consensus, image, etc.

Figure 2.5a[3] shows how both the public and private sector are part of a same system, and how the rules underlying the survival and development of both sectors lie behind their own capability to generate value, to make growth sustainable. This depends on the capability of public and private sector organizations to generate results (e.g. in terms of products, services or rules), which tend to produce an outcome whose value corresponds to an increasing endowment of available resources.

Figure 2.5a also shows how public sector performance does not only feed back under the form of taxes and financial contributions from the community to the benefit of which a given set of services and rules is delivered, but also in terms of external contributions.[4]

So, the private sector feeds back to the public sector: public opinion is primarily affecting the political level, and income primarily affects the funds that the public administration will be able to raise through taxes and other sources, to provide the administrative level with resources to afford public expenditures.

In the described context, a public institution often takes a coordinating role in a system characterized by multiple actors, i.e., public and private institutions. In particular, if we aim to evaluate policy outcomes in such setting, the inter-institutional system's performance would not result from a mere sum of the performance levels produced by each single institution. It would be, rather, the effect of the net relationships and synergies among the different institutions linked to each other.

For instance, to evaluate the outcomes of industrial district policies, a public decision maker (e.g. a regional planner) needs to move the focus of analysis from an institutional to an inter-institutional perspective (Bianchi 2010, pp. 378–381).

---

[3]Though Fig. 2.5 may look like a causal loop diagram, this is not properly the case here. In fact, it tries to capture both the public and private sector into a single and abstract framework. Such framework remarks the role of the public sector into the wider system where it operates, and therefore underlies the main conditions for assessing its performance.

[4]Such additional resources correspond to those that a single public sector institution (or a group of them) ruling a local area is able to procure from third actors (e.g. the Union funding for infrastructure building to the benefit of European Regions). It is worth remarking that this analysis is relevant not only for those public services generating a financial value (e.g. in the case of infrastructures, education, enterprise funding, local area marketing) but also for those generating a qualitative value (e.g. in the case of health care, police or environmental care services, whose indirect outcomes have, however, an economic value too).

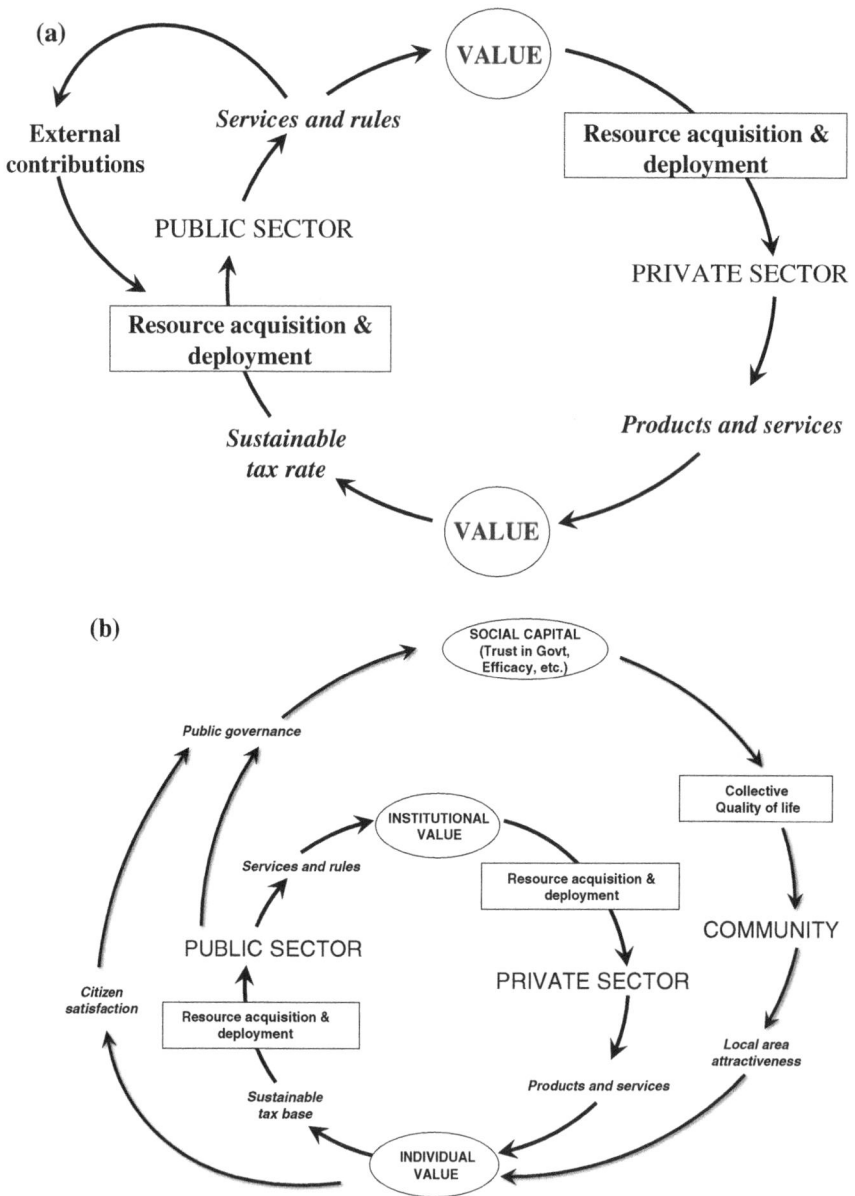

Fig. 2.5 a A systemic framework embodying both the public and private sector: value generation as a focus for assessing performance and a prerequisite for sustaining growth. **b** Combining service delivery improvement with public governance to generate value and trust in government, to deal with 'wicked' problems and to pursue sustainable development of a community's quality of life

2.4.1   Framing Sustainable Growth Within the Inter-Institutional Level (Continued): The Governance of 'Wicked' Problems

The conceptual framework portrayed in Fig. 2.5a advocates the relevance of a value generation and outcome-oriented performance management perspective as a pre-requisite to pursue sustainable growth in a local area. Since the 1980s' and 1990s', this model has inspired most international public sector reforms, aiming to pursue excellent public service delivery as a means to generate value for a community. Though such perspective is still today crucial for the pursuit of social and economic sustainable development, it may not be sufficient. In fact, the dynamic complexity characterizing nowadays' societies is a major cause of amplifying "wicked" problems, whose solutions cannot be found only by service improvement in each of the agencies concerned (Bovaird and Loffler 2003).

"Wicked" problems characterize most of governmental planning, with a specific concern with social issues (Bovaird and Loffler 2007; Rittel and Webber 1973, p. 160). These are complex policy problems featured by high risk and uncertainty and a high interdependency among variables affecting them. "Wicked" problems cannot be clustered within the boundaries of a single organization, or referred to specific administrative levels or ministerial areas. They are characterized by dynamic complexity, involving multi-level, multi-actor and multi-sectoral challenges.

Examples of such problems include social cohesion, climate change, unemployment, crime (Bianchi and Williams 2015), homelessness, healthcare, poverty, education, societal aging (Bianchi 2015), and immigration (Laegreid and Rykkja 2014).

These problems are usually ingrained in major social issues of modern life, whose interpretation is not univocal, because it depends on the adopted value perspectives. Therefore, by simply gathering more information can be insufficient to understand and resolve them. This implies that there is not a definitive (i.e. true or false) solution to them; there can be rather a 'good' or 'bad' way to frame them and to profile one or more consistent (or inconsistent) alternative decision sets (Head and Alford 2013). Wicked problems also imply a multitude of stakeholders. Both the different interests and the multiple mindsets or cultures related to the policy makers who may affect a wicked problem imply that—in order to effectively deal with it—decisions should be made based on a strategic learning process, focused on conflict resolution and dialogue among the players.

Even material and information delays play a major role in characterizing the hidden feedback structure underlying wicked problems' behavior. Therefore, enabling decision makers to promptly perceive weak signals of change and to provide reliable keys to frame them is an important attribute for diagnostic and interactive control systems (Simons 2000, pp. 207–229) in those public sector organizations that should address wicked problems.

Public Administration has always experienced difficulties in dealing with such problems: specifically with respect to its capability to support planning, policy design, decision making, results measurement, assessing policy outcomes, coordinating decision makers and making them accountable to targets. Examples of such difficulties are witnessed by hierarchical forms of organization and systems of control, focused on input monitoring or process compliance, resulting into sharp disconnections between different institutions and among agencies.

Although, since the 1980s, 'New Public Management' (NPM) reforms were designed to fix the described limitations of traditional Public Administration (Meier and Hill 2005, p. 55), their emphasis on decentralization of power has produced unintended effects on the capability of the public sector to affect the outcomes associated with wicked problems. In fact, such reforms have been a major cause of governance fragmentation (Christensen and Laegreid 2007a) and lack of communication in and among agencies.

Policy makers are prone to take symptomatic solutions to wicked problems. For instance, in order to deal with crime, they may focus only input (e.g. police staff) or volume targets (e.g. number of stop-and-frisks), rather than also outcomes; likewise, in order to counteract societal aging, they can be inclined to increase retirement age.

The use of a short-term perspective and a sectoral approach in the formulation and implementation of strategies lead to a static view of the system and to a lack of coordination in policy-making between different public agencies, non-profit and other private stakeholders. This approach may not support governments to identify sustainable actions, whose policy-making usually refers to several jurisdictions, both in terms of level (e.g. national, regional, local) and domain (e.g. policing, welfare, education, justice).

In the last decade, a number of countries have started to develop new approaches that may enable them to improve cohesion, to effectively deal with wicked problems, and pursue a sustainable development of local areas under an inter-institutional perspective. To describe and implement these processes, both the scientific literature and practitioners have coined different terms. Among them are the following: joined-up government[5] (Christensen and Laegreid 2007b, 2013; Christensen et al. 2014),[6] whole-of-government (OECD 2005), integrated governance, outcome steering (Hood 2005), holistic governance, horizontal management (Peters 2015), and new public governance (Osborne 2010).

---

[5]According to Pollitt (2003, p. 35), '"Joined-up government" is a phrase that denotes the aspiration to achieve horizontally and vertically coordinated thinking and action. Through this coordination, it is hoped that a number of benefits can be achieved. First, situations in which different policies undermine each other can be eliminated. Second, better use can be made of scarce resources. Third, synergies may be created through the bringing together of different key stakeholders in a particular policy field or network. Fourth, it becomes possible to offer citizens seamless rather than fragmented access to a set of related services'.

[6]Christensen and Laegreid (2013) describe the Norwegian experience in dealing with such wicked problems, with a specific reference to welfare services. In 2005, Norway merged the central pensions and employment agencies and creates a partnership with locally based welfare services. In the years 2006–2009, Norwegian municipalities established local one-stop-shop welfare offices. In 2008, regional pension units and administrative back offices were established in the counties.

To implement such processes, three main sets of levers should be synergetically managed by governments, that is, (1) institutional reforms, (2) organization structures and performance management systems, and (3) cultural/social systems (Borgonovi 1996, p. 105).

The idea is to design and implement more flexible and pervasive governmental systems that may foster a more pragmatic, less formal and intelligent collaboration among different stakeholders, not only in the public sector sphere.[7]

The implementation of such reforms also implies the use of an outcome-oriented view of performance to frame and assess the desirability of the effects produced by the adopted policies. This approach does not only consider effects in the short run but also in the long run. Furthermore, it does not only focus them in the perspective of a single unit or institution but also under an inter-institutional viewpoint, i.e., that of the relevant system structure generating observed behavior.

By focusing only single (i.e. isolated) input and output measures (e.g. pension or long-term care expenditures, number of retirees, number of working hours), policy makers may be inhibited to assess the aptitude of their own actions to find sustainable solutions that may deal with wicked problems. On the other hand, by combining such measures with outcome performance indicators—for instance related to the community's quality of life (e.g. change in life expectancy from the prevention of unhealthy behavior), labor participation, stakeholders' perceptions and public support to government policies—governments might better assess their own policies' sustainability in both time and space.

A third lever to implement such reforms refers to cultural/social systems. A fundamental change, in terms of cross-sectoral collaboration and coordination, is possible if a strong sense of values, team building, inclusion and trust is fostered among stakeholders. Changing culture and building trust is not an easy and fast process; it requires that a learning-oriented and systems approach can be adopted to support the performance management cycle of each unit.

Figure 2.5b provides a synthesis of the discussion developed so far. It shows that, in order to pursue sustainable growth, 'wicked' problems require that public administrations adopt an outcome-oriented and inter-institutional perspective of performance management, aiming to generate not only value to the benefit of individual institutions (e.g. enterprises or households), but also community value. Such value does not only depend on excellent public service delivery (as remarked in Fig. 2.5a). It also requires that both public and private sector institutions collaborate in public governance, i.e. "the ways in which stakeholders interact with each other in order to influence the outcomes of public policies" (Bovaird and Loffler 2003, p. 316). Therefore, 'good governance' can be meant as "the

---

[7]For instance, in the UK, the Blair government implemented intensive whole-of-government programs, which led to a stronger role of the center (Christensen and Laegreid 2013). Australia and New Zealand governments established new organizational units (e.g. cabinet committees, inter-ministerial or inter-agency collaborative units, inter-governmental councils, task forces, cross-sectoral programs) to foster coordination among different decision makers (Halligan and Adams 2004).

negotiation by all the stakeholders in an issue (or area) of improved public policy outcomes and agreed governance principles, which are both implemented and regularly evaluated by all stakeholders" (Bovaird and Loffler 2003, p. 316).

Likewise good services and rules provided by the public sector may generate new value to the benefit of the private sector (as shown in Fig. 2.5a), good public governance and public service "co-production" (Bovaird 2007) may improve citizens' trust in government and—more broadly speaking—social capital. This is a different kind of public value—in respect to the one that is produced by good public services—since it refers to the quality of community participation to governance and to the ability of a community to generate mutual benefits coming from the aptitude of each player to listen to the others. Such value is a prerequisite to improve a community's quality of life, which is a strategic resource affecting the attractiveness of a local area. Improving the attractiveness of a region may generate further public value, since it may foster more and better economic and social activities, leading to local area sustainable development.

Such perspective of outcome-based performance management in the public sector emphasizes the role of public governance as a means—i.e. as a policy—to improve a community's quality of life and local area attractiveness. In this perspective, both products and services delivered by the private sector (through the support of the public sector) and local area attractiveness (as a local community "product") are able to affect the value each individual can earn in the society, both in financial and non-financial terms. Fostering such value may both increase sustainable taxation and citizen satisfaction and participation to public governance (virtuous sustainable development circle).

2.4.2 Framing Sustainable Growth Within the Inter-Institutional Level (Continued): Financial Restructuring Planning in Local Government

Financial restructuring planning in local government can provide a second example of how an outcome and inter-institutional perspective may foster sustainable development through performance management in the public sector. The topic of financial crisis and growth sustainability in local government has become increasingly significant in the last decade, for Public Administrations. States are struggling to balance objectives such as: GDP growth, employment, quality of life, and financial equilibrium. Municipal bankruptcies and the need to outline sustainable restructuring plans are today crucial in many countries of the world.

The many targets Public Administrations attempt to address may look as diverging from each other, if framed on a static and sectoral perspective. Such view is often adopted by both law prescriptions and the professional practice. Often the primary focus of analysis is on only financial statements and on the adjustment of debts.

In such perspective, the search for the causes behind the crises is primarily done in financial and juridical terms, strictly related to the institutional dimension of the perceived problems. Such approach may not support policy makers to understand how social and managerial phenomena in the wider system have affected the current insolvency state of an institution. Therefore, it may not support them to outline sustainable policies in the long run, to counteract the financial crises at their own roots.

Most of such factors can be outside the institutional domain and the control of local government. Examples of such factors are: the rates at which the stocks of population and enterprises change over time, employment rates, crime rates, perceived public service levels (e.g. health care, education, transportation), image and attractiveness of an urban area, trust and loyalty towards government, social capital, quality of infrastructures, capability to network with other public and private sector organizations. Public policy makers often misperceive even the delays affecting the accumulation and depletion processes of local strategic resources. The endowment and deployment of the strategic resources in a region may differently affect the drivers of unsuccessful financial results.

Therefore, the identification of the leverage points on which to act in order to design and implement sustainable restructuring plans in local governments should go beyond the financial and the institutional dimensions (e.g. increasing tax rates, selling property, negotiating new loans, or bargaining debts maturity extensions).

In order to recover financial equilibrium and competitiveness, many New Public Management (NPM) reforms have been characterized by a sectoral and too partial approach. If one considers the current practice in local strategic planning, one may perceive how urban planners can be inclined to over-emphasize the architectural and land-use perspectives associated with the development of metropolitan plans; sociologists may devote more attention to the effects of group behavior and culture on local performance; accountants and financial experts may be too focused on the technical aspects related to the drawing up of budgets and reports, often linked to the formal procedures through which public sector decision makers are legitimated to obtain the resources to implement policies; experts in regional studies may over consider macro-economic aggregates (e.g. consumption rates, savings, employment); political scientists and lawyers may overweigh the role of rules and formal institutional systems.

Though the viewpoint of each discipline may be considered as consistent with the analyzed topic—if observed within the framework of a specific study-area—a sectoral approach runs the risk not to be able to capture the systemic, complex and dynamic structure of the problem context. Therefore, an inter-disciplinary and learning-oriented system perspective is needed.

There is a gap, in both professional and organizational terms, in today's Public Administration between its current and expected capability to deal with dynamic complexity. The use of SD modeling may significantly enhance the capability of governments to fill such gap. SD can successfully support the drawing up of restructuring and reorganization plans, at both local and central government, to outline sustainable policies that look beyond a 'debt adjustment' perspective, and may support decision makers' learning processes about a local area's capability to build up and preserve a sustainable competitive advantage and community development.

2.5 Fostering Sustainable Organizational Development: From Balanced Scorecards to Dynamic Performance Management Systems

From the previous discussion, it is possible to observe two issues.

First, though different organizations may sharply differ from each other because of various structural factors, the same conceptual framework should support their own performance management cycle. In fact, performance should be evaluated according to the aptitude of an organization to pursue a growth rate that balances the short with the long term and is also consistent with the physiological goals of the organization. These goals compose elements in a wider socio-economic system to which an organization belongs and to whose continuity and sustainable growth it must therefore contribute.

And second, the current knowledge and practice in strategic planning and performance management are not able properly to deal with sustainable growth.

In particular, conventional financially-focused P&C systems have been considered lacking in relevance (Johnson and Kaplan 1987; Kaplan and Norton 1996), since they are not able to provide information that can support either dynamic complexity management, the measurement of intangibles, the detection of delays, adequate understanding of the linkages between the short and the long term, and the setting of proper system boundaries in strategic planning.

To cope with such problems, the Balanced Scorecard (BSC) has been used by many organizations both in the private and public/non-profit sectors. The two main concepts underlying the BSC framework are:

1. Organizational performance cannot be managed by focusing only on end-results: one should understand how such results are generated, which factors affect them, and how decision-makers can be made accountable for them.
2. Performance cannot be gauged only in terms of financial measures. Also, a "customer", a "process", and a "learning and growth" perspective is needed. These three additional BSC perspectives on performance may allow one to understand to what extent financial performance is sustainable in both time and space.

According to Kaplan and Norton, the BSC enables companies to measure financial results while simultaneously monitoring progress in building capabilities and acquiring the intangible assets they need for future growth (Kaplan and Norton 1996). Therefore, they explicitly recognize the BSC as a strategic tool for the control of both lag and lead indicators (Norton 2001, p. 4).

The increasing popularity of the BSC is due to the support it gives to management in avoiding disconnections between strategy and implementation. The BSC also stresses the idea of cause-and-effect relationships between measures, in order to avoid the possibility that performance improvement in one area may be at the expense of performance in other areas. Kaplan and Norton, indeed, explicitly stated the systemic interrelationships within and between the four key BSC perspectives, incorporating both lead and lag indicators, which impact on organizational performance (Martinsons et al. 1999, p. 83).

This approach aims at offering a systematic and comprehensive road map for organizations to follow in translating their mission statements into a coherent set of performance measures. These measures are not only intended to control company performance, but also to articulate and communicate the organization's strategy (Mooraj et al. 1999, p. 490) and to help align actions from different levels of management for the achievement of a common goal (Malina and Selto 2001, p. 54).

Furthermore, the BSC enhances managers' understanding of strategies and stimulates the creation of a common company vision. The BSC, indeed, forces managers to elicit, compare and discuss their implicit assumptions and beliefs and to articulate them for the formulation of company's strategy (Malmi 2001, p. 210–214). In fact, managers are requested to contribute to the implementation of the BSC by identifying a set of objectives that are connected by causal relationships that are consistent with the vision and mission of the company.

However, it has been remarked how—in order to encourage openness and frankness of expression (Wisniewski and Dickson 2001, p. 1065)—the support of an external facilitator leading the BSC construction process is often necessary. This would also allow the elicitation of managers' mental models.

In spite of its widely recognized advantages, even the BSC presents certain conceptual and structural shortcomings. Linard et al. (2002) assert that the BSC fails to translate company strategy into a coherent set of measures and objectives, because it lacks a rigorous methodology for selecting metrics and for establishing the relationship between metrics and corporate strategy.

Sloper et al. (1999) remark that the BSC is a static approach. Although Norton and Kaplan stress the importance of feedback relationships between BSC variables for describing the trajectory of a given strategy, the cause-and-effect chain is always conceived as a bottom-up causality, which totally ignores feedbacks, thereby confining attention only on the effect of variables in the lower perspectives (Linard and Dvorsky 2001).

In particular, the BSC approach does not help one to understand:

- How strategic resources accumulation and depletion processes triggered by the use of different policy levers affect performance drivers;
- How performance drivers affect end-results (both output and outcome measures);
- How end-results will affect strategic resource accumulation and depletion processes.

In order to provide decision-makers with proper lenses for interpreting such phenomena, understanding the feedback loop structure underlying performance, and identifying alternative strategies to adopt so as to change the structure for performance improvement, SD modeling has been used (Bianchi and Montemaggiore 2008; Kaplan and Norton 2001, pp. 311–313; Linard and Yoon 2000; Morecroft 2007; Richmond 2001; Ritchie-Dunham 2001; Warren 2008). SD models can be properly linked to either accounting or financial models to support strategic planning and control (Bianchi 2002) and, also, to implement dynamic performance management.

This topic will be illustrated in the next chapter of this book.

References

Ackoff, R. (1986). *Management in small doses*. New York: Wiley.

Bianchi, C. (2002). Introducing SD modeling into planning and control systems to manage SMEs' growth: A learning-oriented perspective. *System Dynamics Review, 18*(3), 315–338.

Bianchi, C. (2010). Improving performance and fostering accountability in the public sector through system dynamic modeling: From and 'external' to an 'internal' perspective. *Systems Research and Behavioral Science, 27*, 361–384.

Bianchi, C. (2015). Enhancing joined-up government and outcome-based performance management through system dynamics modeling to deal with wicked problems: The case of societal ageing. *Systems Research & Behavioural Science, 32*, 502–505.

Bianchi, C., & Montemaggiore, G. (2008). Enhancing strategy design and planning in public utilities through "dynamic" balanced scorecards: Insights from a project in a city water company. *System Dynamics Review, 24*(2), 175–213.

Bianchi, C., & Rivenbark, W. (2012). A comparative analysis of performance management systems. The cases of Sicily and North Carolina. *Public Performance & Management Review, 35*(3), 509–526.

Bianchi, C., & Williams, D. (2015). Applying system dynamics modeling to foster a cause-and-effect perspective in dealing with behavioral distortions associated with a city's performance measurement programs. *Public Performance Management Review, 38*, 395–425.

Bianchi, C., & Xavier, J. A. (2014). The design and execution of performance management systems at state level: A comparative analysis of Italy and Malaysia. *APPAM Fall Research Conference*. November 6–8, Albuquerque.

Bovaird, T. (2007). Beyond engagement and participation: User and community coproduction of public services. *Public Administration Review, 5*, 846–860.

Bovaird, T., & Loffler, E. (2003). Evaluating the quality of public governance: Indicators, models and methodologies. *International Review of Administrative Sciences, 69*(3), 313–328.

Bovaird, T., & Loffler, E. (2007). Assessing the quality of local governance: A case study of public services. *Public Money & Management, 27*(4), 293–300.

Borgonovi, E. (1996). *Principi e sistemi aziendali per le Amministrazioni Pubbliche (Management Principles and systems for Public Administrations)*. Milano: Egea.

Christensen, T., Fimreite, A. L., & Lægreid, P. (2014). Joined-up government for welfare administration reform in Norway. *Public Organization Review, 14*(4), 439–456.

Christensen, T., & Laegreid, P. (2007a). *Transcending new public management*. Surrey: Ashgate.

Christensen, T., & Laegreid, P. (2007b). The whole-of-government approach to public sector reform. *Public Administration Review, 67*(6), 1059–1066.

Christensen, T., & Laegreid, P. (2013). Welfare administration reform between coordination and specialization. *International Journal of Public Administration, 36*, 556–566.

Coda, V. (2010). Entrepreneurial values and strategic management. Essays in management theory. New York: Palgrave MacMillan.

Dyson, R. (2000). Strategy, performance and operational research. *Journal of the Operational Research Society, 51*(1), 5–11.

Flamholtz, E. (1996). Effective organizational control. A framework, applications, and implications. *European Management Journal, 14*(6), 596–611.

Flamholtz, E., & Hua, W. (2002) Strategic organizational development, growing pains and corporate financial performance: An empirical test. *European Management Journal, 20*(5), 527–536.

Greiner, L. E. (1972). Evolution and revolution as organizations grow. *Harvard Business Review, 50,* 4.

Halligan, J., & Adams, J. (2004). Security, capacity and post-market reforms, public management change in 2003. *Australian Journal of Public Administration, 63*(1), 85–93.

Hamel, G., & Prahalad, C. (1994). Competing for the future. Boston: Harvard Business School Press.

Head, B., & Alford, J. (2013). Wicked problems: Implications for public policy and management. *Administration & Society,* Published online, doi:10.1177/0095399713481601

Henri, J. F. (2006). Organizational culture and performance management systems. *Accounting, Organizations and Society, 31*(1), 77–103.

Hood, C. (2005). The idea of joined-up government. A historical perspective. In Bogdanor V. (Ed.), *Joined-up government.* Oxford, British Academy: Oxford University Press.

Kaplan, R., & Norton, D. (1996). *The Balanced scorecard: Translating strategy into action.* Boston: Harvard Business School Press.

Kaplan, R., & Norton, D. (2001). The strategy-focused organization. Boston: Harvard Business School Press.

Kloot, L. (1997). Organizational learning and management control systems. *Management Accounting Research, 8,* 47–73.

Johnson, H., & Kaplan, R. (1987). The rise and fall of management accounting. Mass. Boston: Harvard Business School Press.

Laegreid, P., & Rykkja, L. H. (2014). Governance for complexity—How to organize for the handling of «Wicked Issues»? *Stein Rokkan Centre for Social Studies.* https://bora.uib.no/handle/1956/9384

Langfield-Smith, K. (1997). Management control systems and strategy: A critical review. *Accounting, Organizations and Society, 22*(2), 207–232.

Linard, K., & Dvorsky, L. (2001) People—Not human resources: The system dynamics of human capital accounting. In *Operations Research Society Conference.* Bath: University of Bath.

Linard, K., Fleming, C., & Dvorsky, L. (2002) System dynamics as the link between corporate vision and key performance indicators. In *Proceedings of the 20th System Dynamics International Conference.* Palermo: System Dynamics Society.

Linard, K., & Yoon, J. (2000). The dynamics of organizational performance development of a dynamic balanced scorecard. In *The First International Conference of System Thinking in Management* (pp. 359–364).

Lorange, P., & Vancil, R. (1976) How to design a strategic planning system. *Harvard Business Review, 54,* 75–81.

Malina, M. A., & Selto, F. H. (2001). Communicating and controlling strategy: An empirical study of the effectiveness of the balanced scorecard. *Journal of Managerial Accounting Research, 13,* 47–90.

Malmi, T. (2001). Balanced scorecards in Finnish companies: A research note. *Management Accounting Research, 12,* 207–220.

Martinsons, M., Davison, R., & Tse, D. (1999). The balanced scorecard: A foundation for the strategic management of information systems. *Decision Support Systems, 25*(1), 71–88.

Meier, K. J., & Hill, G. (2005). Bureaucracy in the twenty-first century. In: E. Ferlie, L. Lynn & C. Pollitt (Eds.), *The Oxford handbook of public management* (pp. 51–71). Oxford: Oxford University Press.

Mintzberg, H., & Westley, F. (1992). Cycles of organizational change. *Strategic Management Journal, 13,* 39–59.

Moore, M. (1995). *Creating public value. Strategic management in government.* Cambridge, Mass: Harvard University Press.

Mooraj, S., Oyon, D., & Hostettler, D. (1999). The balanced scorecard: A necessary good or an unnecessary evil? *European Management Journal, 17*(5), 481–491.

Morecroft, J. (2007). *Strategic modeling and business dynamics.* Chichester: Wiley.

Munro, M., & Wheeler, B. (1980). Planning, critical success factors, and management's information requirements. *Management Information Systems Quarterly, 4*(4), 27–38.

Norton, D. (2001). Building strategy maps. Part 3: The importance of time-phasing the strategy. In *Balanced scorecard: Insight, experience and ideas for strategy focused organizations* (pp. 1–4). USA: Harvard Business School. March/April.

OECD. (2005). Modernizing government. Paris: The Way Forward. http://www.oecd.org/gov/modernisinggovernmentthewayforward.htm

Osborne, S. P. (Ed.) (2010). The new public governance: Emerging perspectives on the theory and practice of public governance. London: Routledge.

Otley, D. (1994). Management control in contemporary organizations: Towards a wider framework. *Management Accounting Research, 5,* 289–299.

Peters, G. (2015). Pursuing horizontal management. The politics of public sector coordination. Lawrence, Kansas: University Press of Kansas.

Pollitt, C. (2003). Joined-up government: A survey. *Political Studies Review, 1,* 34–49.

Putnam, R. (2000). *Bowling alone: The collapse and revival of American community.* New York: Simon and Schuster

Radermacher, W. (1999). Indicators, green accounting and environment statistics—Information requirements for sustainable development. *International Statistical Review, 67*(3), 339–354.

Rainey, H. G., & Han Chun, Y. (2005). Public and private management compared. In E. Ferlie, L. E. Lynn, & C. Pollitt (Eds.), *The oxford handbook of public management.* Oxford: Oxford University Press.

Riccaboni, A. & Leone, E. L. (2010). Implementing strategies through management control systems: The case of sustainability. *Journal of Productivity and Performance Management, 59* (2), 130–144.

Richmond, B. (2001). A new language for leveraging scorecard-driven learning. Reprinted from "Balanced Scorecard Report". *Harvard Business School Publishing, 3*(1(), 11–14.

Ritchie-Dunham J. L. (2001). Informing mental models for strategic decision making with ERPs and the balanced scorecard: A simulation-based experiment. In *Proceedings of the 19th International System Dynamics Conference.* Atlanta: System Dynamics Society.

Rittel, H. W. J., & Webber, M. M. (1973). Dilemmas in a general theory of planning. *Policy Sciences, 4,* 155–169.

Schreyögg, G., & Steinmann, H. (1987). Strategic control: A new perspective. *The Academy of Management Review, 12*(1), 91–103.

Simons, R. (2000). *Performance measurement & control systems for implementing strategy.* New Jersey: Prentice Hall, Upper Saddle River.

Sloper, P., Linard, K., & Paterson, D. (1999). Towards a dynamic feedback framework for public sector performance management. In *Proceedings of the 17th International System Dynamics Conference,* Wellington: System Dynamics Society.

Sorci, C. (2007). *Lo sviluppo integrale dell'azienda (Holistic Business Growth).* Milan: Giuffrè.

Talbot, C. (2005). Performance management. In: E. Ferlie, L. E. Lynn & C. Pollitt (Eds.), *The Oxford Handbook of Public Management.* Oxford: Oxford University Press.

Warren, K. (2008). *Strategic management dynamics.* New York: Wiley.

Wernerfelt, B. (1984). A resource-based view of the firm. *Strategic Management Journal, 5*(2), 171–180.

Werther, W. B., & Chandler D. B. (2006). *Strategic corporate social responsibility.* Thousand Oaks: Sage.

Wisniewski, M., & Dickson, A. (2001). Measuring performance in dumfries and galloway constabulary with the balanced scorecard. *Journal of the Operational Research Society, 52,* 1057–1066.

Chapter 3
Fostering Sustainable Organizational Development Through Dynamic Performance Management

3.1 Designing Dynamic Performance Management Systems to Enhance Sustainable Organizational Development: Three Complementary Views

This chapter outlines a conceptual and methodological framework to support organizational policy makers in managing and assessing performance within the perspective of sustainability. It analyzes the role of SD to enhance P&C systems so as to both manage performance and foster sustainable development.

The need for a dynamic performance management (DPM) approach is discussed. To this end, the concept of balanced growth and the institutional and inter-institutional levels of analysis that have been discussed in the previous chapter will be now framed through three interconnected views of organizational performance, i.e.: instrumental, objective, and subjective. Applications of such framework, based on case-study analysis, are also illustrated.

Designing a P&C system to support decision makers to assess performance under a sustainability perspective is the core of DPM. It requires a selective and sequential method of inquiry.

DPM is an approach that enables organization decision makers to frame the causal mechanisms affecting organizational results over time. Such a field of research and practice is based on two converging methods of inquiry: Performance Management and SD modeling.

DPM takes its own premises from the literature that has demonstrated the lack of relevance of conventional financially focused P&C systems. Such systems are no longer able to provide information that can support: the management of dynamic complexity, measurement of intangibles, detection of delays, understanding linkages between short- and long-term, and setting proper system boundaries in strategic planning. In order to cope with such problems, to provide decision-makers with proper lenses to interpret such phenomena, to understand the feedback structure underlying performance, and to identify alternative strategies to change

© Springer International Publishing Switzerland 2016
C. Bianchi, *Dynamic Performance Management*, System Dynamics
for Performance Management 1, DOI 10.1007/978-3-319-31845-5_3

the structure for performance improvement, SD modeling has been used to support an understanding of: (1) how end-results can be affected by performance drivers; (2) how performance drivers can, in turn, be affected by the use of policy levers aimed to influence strategic resource accumulation and depletion processes; and (3) how the flows of strategic resources are affected by end-results.

3.2 The Instrumental View of Performance

In order to apply DPM as an approach to enhance sustainable organizational development, the "instrumental" view must be taken first on a corporate or divisional level, or both. This level of investigation allows one to frame a synthetic picture of the key performance factors for the overall organization. Although it can be considered as only an initial step, based on which a more detailed inquiry can be later carried out on a departmental level, the results it delivers can effectively introduce DPM in an organization whose culture or size may not be ready to sustain a more challenging and pervasive investigation. Depending upon the cultural level, size, available time and other resources of the client organization, this level of study can also be conducted through insight stock-and-flow DPM modeling or even qualitative mapping, based on DPM charts.

The "instrumental" view implies that alternative means for improving performance be made explicit. In this regard, it is necessary to identify both end-results and their respective drivers. To affect such drivers, each responsibility area must build up, preserve, and deploy a proper endowment of strategic resources that are systemically linked to each other.

Figure 3.1 illustrates how the end-results provide an endogenous source inside an organization for the accumulation and depletion processes that affect strategic resources. In fact, they can be modeled as in or out-flows, which over a given time span change the corresponding stocks of strategic resources, as the result of actions implemented by decision-makers. For instance, liquidity (a strategic resource) may change as an effect of cash flows (an end-result); the image and credibility of an organization towards citizens (strategic resources) may change as an effect of changes in their satisfaction levels (an end-result).[1] There also are interdependencies between different strategic resources: image may affect the capability of an organization to get funds from different stakeholders. Furthermore, both image and financial resources may affect an organization's capability to recruit skilled human resources and keep them.

So, each strategic resource should provide the basis to sustain others in the same system. For instance, in a Municipality, both workers and equipment provide

---

[1]An interesting systemic-control model based on a different perspective, in respect to the instrumental view here discussed, has been proposed by Schwaninger (2009, p. 48).

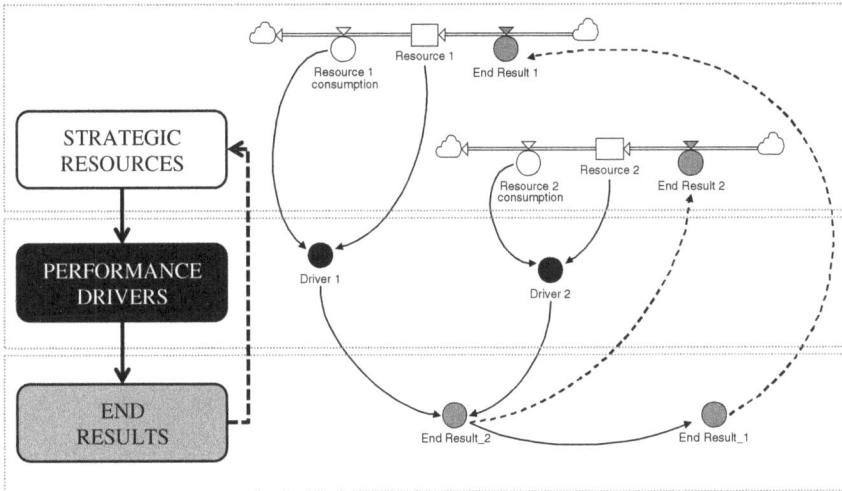

Fig. 3.1 The "instrumental" view of performance

capacity, which influences the perceived service quality. This affects regional attractiveness, which, in turn, influences population dynamics. A change in regional population will affect workload and perhaps the stock of available financial resources, and eventually capacity and service. The feedback loops underlying the dynamics of the different strategic resources imply that the flows affecting such resources are measured over a time lag. Therefore, understanding how delays influence strategic resources and achieved results becomes a key issue for managing performance in dynamic complex systems.

Organizational growth can be sustainable if the rate at which end-results change the endowment of corresponding strategic resources remains balanced. This implies that management is able to:

- Gradually increase the mix of strategic resources, and not only a bounded group of them, e.g., when a company fosters commercial growth only by expanding its sales force and other commercial strategic resources, without gradually increasing its production capacity. This concept leads to the institutional view of growth, which was analyzed in the previous chapter.
- Resource increase is not obtained by reducing the endowment of the wider strategic resources in the local area or industry. This concept leads to the inter-institutional view of growth, which was analyzed in the previous chapter.

End-results can be measured over a sequential chain and positioned on several layers (Fig. 3.2).

End-results on the first layer are those that most synthetically measure the overall organizational performance. They are flows that change the endowment of strategic resources that cannot be purchased in the market, i.e.:

Fig. 3.2 Further layers
affecting first layer end-results

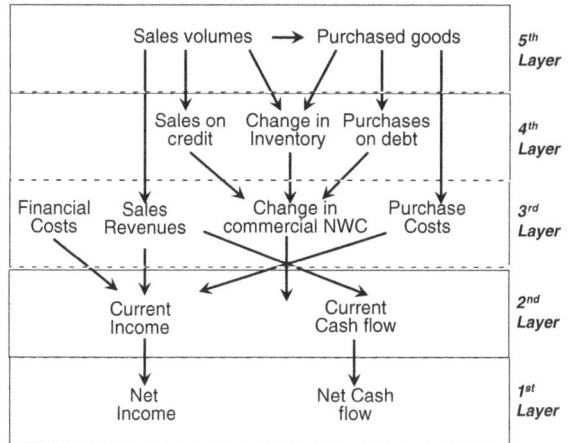

1. *Liquidity and equity.* Though both resources can also be negotiated through
   loans and the issue of shares, respectively, a sustainable development implies
   that an organization is able to increase them through its cash flows and profits;
2. Resources *generated by management (internal) routines.*[2]

An organization can purchase *physical resources* (e.g. inventory, labor), or
production *capacity* (e.g. related to people, machinery, or computers). It can
alternatively negotiate contracts for their service use. It can also purchase or build
information resources. On the other hand, it cannot "purchase" stakeholders' perceptions, customer base, product portfolio, and—to some extent—even commercial
networks. In fact, though the property of patents and product models, or a branch
network can be purchased, only the organization's capability to build on such
property and add value on it may allow a business to retain and perhaps to attract
customers.

So, end-results affecting *resources generated by management (internal) routines*
refer to intangible resources that can be associated with: (a) perceptions that
stakeholders in the competitive and social system have about the organization. Such
perceptions (e.g. regarding its reputation and solvency) can change because of the
overall performance—i.e. the organization's end-results; (b) any other strategic
asset an organization cannot purchase through market negotiations, since they can
be built from only inside the business (e.g. organizational climate, employees'
burnout, morale, intellectual capital, capabilities, product quality, customer base,
and product portfolio).

---

[2]Winter (1964, p. 263) defined a routine as "pattern of behavior that is followed repeatedly, but is
subject to change if conditions change".

Fig. 3.3 End-results
affecting strategic resources

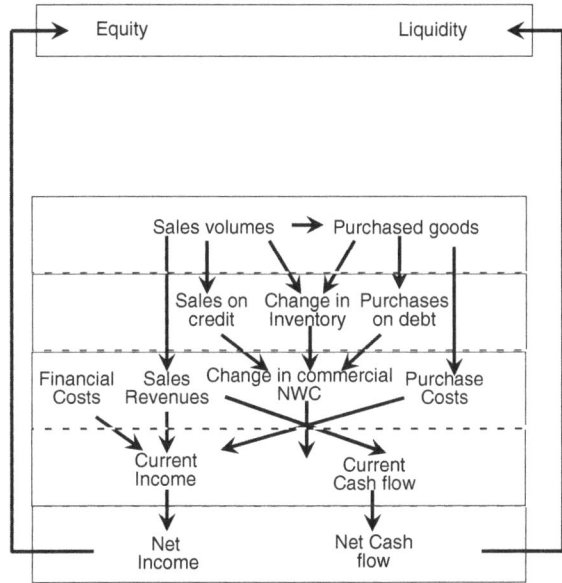

To affect the results positioned on this first layer, further layers must be identified.

For example, net income and cash flows can be affected by current income and the current cash flow, respectively. Such flows can be further affected by third-layer end-results, i.e., sales revenues, financial costs, purchase costs, and the commercial net working capital flow. These more detailed financial measures are, in turn, affected by non-monetary end-results. So, sales volumes affect sales revenues and, through sales on credit and changes in inventory, also the net working capital flow. They also affect purchase volumes, which impact on purchase costs and, through purchases on credit and changes in inventory, on the net working capital flow. Therefore, in this example, sales volumes can be located on a last layer of end-results.

Figure 3.3 shows the relationship between end-results and strategic resources, related to the example here illustrated.

End-results should both include *output* and *outcome* measures. Since both kinds of indicators are flow variables, their units of measure are defined as a ratio between: (1) the measure of the affected strategic resource (stock), and (2) time. Though the metrics for measuring them is the same, there are strong differences between the two of them.

Output measures are workload (or volume) indicators (Ammons 2001, pp. 12–14), e.g.: number of processed applications in the last week, number of students who passed an exam in the last semester, number of arrests performed by police in a month, collected garbage or number of trees planted by a Municipality in a quarter, production or shipping rates, sales or purchase orders rates, sales revenue rates, purchase costs rates.

Outcome measures depict the aptitude of the recorded outputs to: provide the users with the desired service levels (e.g.: quality, time, scope, and price), or to generate a change in the endowment of strategic resources shared by different institutions in a region. Also the previously said *first layer* end-results—affecting both liquidity and equity, and strategic resources generated by *management routines*—are outcome measures.

Though both typologies of performance indicators are end-results for an organization, outcome measures imply that a longer time horizon and broader system boundaries are adopted to measure and manage them, in respect to output indicators.

Lack of focus on both kinds of measures runs the risk of establishing myopic P&C systems, i.e. that performance management systems may not be able to support organizations to manage their own sustainable development.

End-results can be affected through performance drivers. Both end-results and performance drivers are needed in a DPM system: while the former can be influenced in the medium/long-run, the latter can be influenced in the short run. They provide decision makers with measures of possible weak signals of future change in the end-results: this allows an organization to perceive and measure the effects of discontinuity on performance, and perhaps to counteract them. Therefore, such measures can be useful in order to envisage possible changes in the financial, competitive, or social end-results. For instance, a decline in a competitive performance driver might suggest that future sales orders, revenues, profits and cash flows might decline as well, in spite of a strong current financial position of the organization. Likewise, an improvement in a performance driver (or in a set of them) might suggest that a restructuring policy is producing benefits that will allow the organization to recover the targeted end-result flows in the future.

Competitive performance drivers are associated with critical success factors in a strategic business area. They can be measured in relative terms, i.e., as a ratio between the organizational performance perceived by clients and a benchmark or target.

Such denominators are usually defined as standard (or normal or desired) values. They must be gauged in relation to either perceived past performance, or clients' expectations, or competitors' performance, or even as a function of the achievable targets by the organization (budget values).

Competitive performance drivers can be related to any business functional area, e.g.: R&D, production, commercial, etc.

R&D drivers refer to the acquisition of a competitive advantage, e.g. in relation to:

- Quality (e.g.: rework, detected errors),
- Time, and
- Volume.

Production drivers can be related to factors, e.g. in relation to:

- Quality. Examples are: (1) *Project quality* (% features fitting with project specifications); (2) *Quality conformity*, (defective products %, rework time %, number of post-sale product faults); (3) *Product reliability and maintainability* (number of post-warranty 'rescue' visits to customers, mean time between failure, average product repair time); (4) *Support service* in product installation and use (average time between customer call for assistance and service delivery).
- Flexibility. Examples are: (1) *Volume flexibility* (aptitude to promptly change production quantity); (2) *Product mix flexibility* (aptitude to promptly change production mix); (3) *Product scope and time to launch* new *products*; (4) *Promptness* (time to address customer requests, reliability of company response time, aptitude to respond to sudden changes in purchase orders);
- Service. Examples are: (1) *support and post-sale assistance*, and (2) *product customization*.
- Efficiency, i.e., the aptitude to saturate resource capacity and to maximize throughput.

Examples of commercial drivers are: delivery delay, shipping rate, minimum sales order, price, advertising spend, market coverage,[3] physical distribution proximity to clients.

Also *social performance drivers* can be measured in terms of ratios between company strategic assets and a target, which can mostly be expressed in terms of either stakeholder's expectations or perceived past organizational performance. For example, a social performance driver could be referred to the ratio between the actual and planned number of perceived social initiatives undertaken by a firm.

Financial performance drivers also can be measured in relative terms. For instance, the debts-to-total investments ratio often affects the change in company solvency perceived by funders. Such driver is the ratio between two stocks. Efficiency measures affecting operational costs can be gauged in terms of ratios as well. For example, the employee's time per unit of workload is an expression of the ratio between two stocks—employees (unit of measure: people) and workload (unit of measure: widgets per week), multiplied by a constant (working hours per people per week).

Among financial performance drivers, the cost accounting literature (Cooper and Kaplan 1988) has devoted a specific attention to *cost drivers*. They are related to factors generating cost variability, due the carried out processes. A cause-and-effect relationship exists between the absorption of a production factor in the fulfillment of processes and the sustained costs.

Two main kinds of cost drivers can be identified:

- Volume (or parametric) cost drivers, and
- Transaction cost drivers.

---

[3]I.e., the ratio between the company sales agents and the potential customers.

While "parametric" cost drivers (e.g. the unit consumption rate of raw material per widget produced) underlie a proportional relationship between the use of a resource and its cost—as a function of activity volumes—"transaction" cost drivers do not affect costs related to aggregated production volumes. They rather influence costs related to the complexity that characterizes those organizations competing based on their own capability to address a wide range of users and needs. Transaction cost drivers can be related to:

- R&D and Production: e.g., number of runs/shifts and associated set-up costs; number of engineering change orders and related product design costs; number of products/models and related scheduling costs.
- Quality control: e.g., number of production batches or processes and associated sample-testing costs.
- Distribution: e.g., number of stock keeping units and related warehousing and transportation costs.
- Administration: e.g., number of customers/geographic zones of sales or number of sales orders, or invoices, and related revenue cycle costs; and also: number of purchase orders, with the related material inspection and expenditure cycle costs).

If framed in a broader context than cost accounting, transaction cost drivers may provide decision makers with useful insights for strategic analysis, such as the possibility to frame:

- Trade-offs in time. For instance: higher set-up costs and capacity cost savings (in the short run) versus technology investments/production flexibility (in the long run).
- Trade-offs in space (and time). For instance: transaction costs savings (in the short run) versus higher customer service in the long run; cost leadership versus differentiation/service competitive advantage.

Figure 3.4 shows an example where competitive performance drivers are "product delivery delay" and "product scope" (e.g. the number of product models in a catalogue). A variable named "competitive performance" is also set as a synthetic performance driver combining both "delivery delay" and "product scope".[4]

The figure also shows two main financial performance drivers, i.e.: the days of sales outstanding (DSO) and the average inventory time. They, respectively, affect sales volumes and financial costs.

Figure 3.5 shows an example of how policies aimed to leverage on strategic resources may influence competitive performance drivers.

---

[4]Since both are normalized (i.e. relative) measures, "competitive performance" would be a weighted average between the two drivers, where the weight attributed to each of them is a measure of its contribution to the competitive position of the firm.

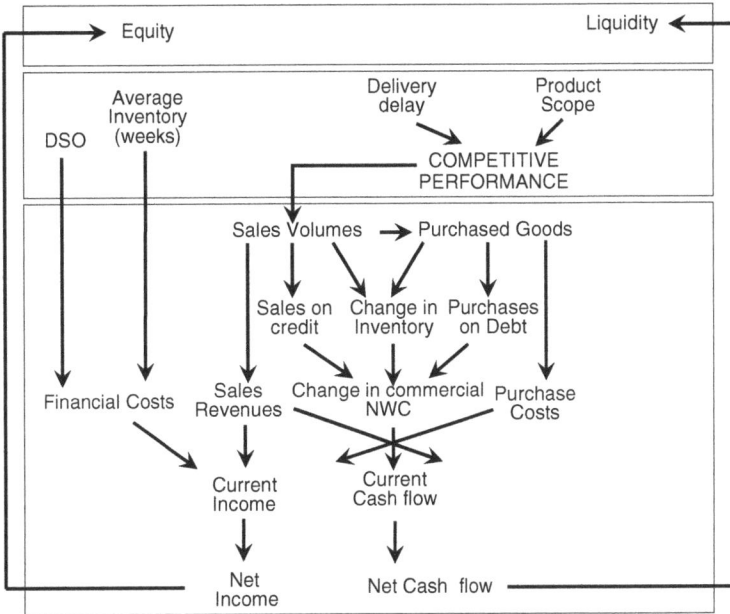

Fig. 3.4 Competitive and financial performance drivers affecting end-results

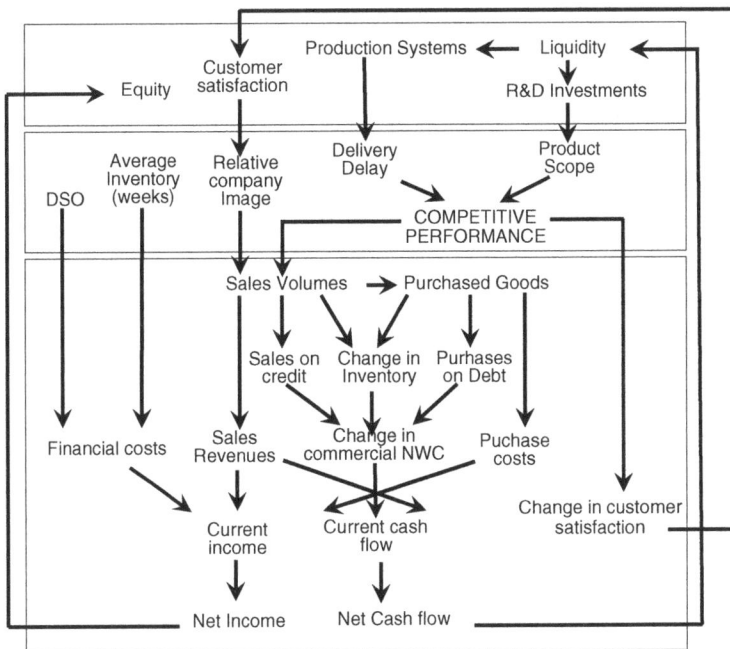

Fig. 3.5 Strategic resources affecting performance drivers

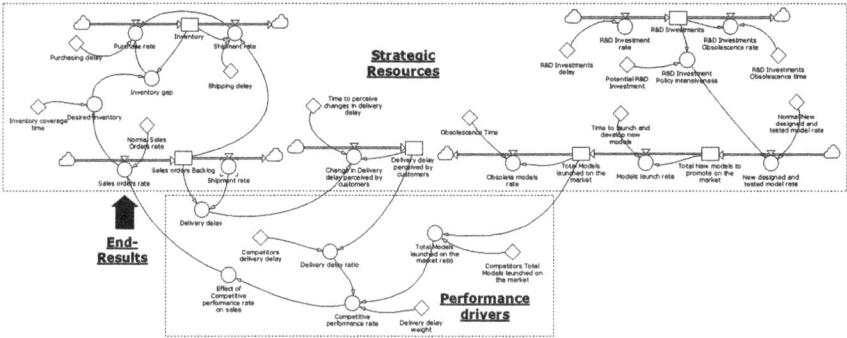

Fig. 3.6 A simplified version of a system dynamics stock-and-flow model mapping competitive performance according to the "instrumental" view

More specifically, the figure illustrates how delivery delay and product scope are, respectively, affected by investments in production systems and R&D. Such investments are, in turn, influenced by company liquidity.

Furthermore, the figure shows how the identification of strategic resources may allow one to detect further performance drivers. In fact, the relative company image (e.g., in respect to competitor's image) can be considered as a further variable driving sales volumes. A better image, in respect to competitors, would act as a multiplier of the sales orders and volumes that a company might get because of its current delivery delay and product scope performance. Figure 3.5 also shows that the customer satisfaction level (strategic resource) affects company image. The change in customer satisfaction identifies another competitive end-result, which is in turn affected by competitive performance.

Figure 3.6 portrays an example of a stock-and-flow model mapping competitive performance. It shows how a business strategic resource system affects two competitive performance drivers, i.e., the "product launched on the market ratio" and "delivery delay ratio." The first driver is a ratio between the "product launched on the market" and a benchmark, which is modeled through an input parameter, i.e., the "competitors' products launched on the market." The second driver is a ratio between the "delivery delay perceived by customers" and another benchmark, i.e., the "competitors' delivery delay."

Both drivers are weighted according to their relative importance in affecting customer demand. So, the "competitive performance ratio" is a synthetic measure of the previous drivers. This affects a multiplier, i.e., a normalized graph function, named "effect of competitive performance ratio on sales." This is multiplied by the normal sales order rate (an input parameter) to determine the sales orders rate (an end-result). The sales rate affects both desired inventory and the purchase order rate.

Inventory levels provide shipments, which allow the firm to manage the delivery delay. Likewise, R&D investments affect the R&D policy intensiveness, which in turn affects new-models development and product launching.

3.3 Operationalizing the Instrumental View. From Static to Dynamic Performance Measures: A Shift of Mind

The analysis that has been carried out so far has emphasized how both end-results (flow variables) and performance drivers (auxiliary ratio variables) provide organizations with measures on which to focus planning and reporting.

A shift of mind is needed when a dynamic view of performance management is adopted. There are common errors and misunderstandings that may happen when this approach is introduced to people who are used to static performance management.

One of the main problems that tackle such a shift of mind is the lack of aptitude by organizations to perceive the *relationship between means and ends*. Static performance management tends to overlap the two things.

3.3.1 Resource Measures and Performance Measures

Resource measures should not be confused with *performance measures*. There is a *means* versus *ends* relationship between the two. If a static view of performance management is adopted, decision makers may not be enabled to frame such nexus. In a static perspective, strategic plans run the risk to be a list of abstract and non-measureable goals (in respect to the outcomes they should underlie), which are disconnected with management objectives. On the other side, departmental budgets and actions plans run the risk to be a description of activities.

By following this approach, one may include among the aggregate departmental performance measures generic objectives related to the fulfillment of projects by the lower responsibility levels. For example, a public sector organization which embodies in a departmental budget targets such as "launching a website" or "accomplishing the highway XYZ"—measured in terms of project accomplishment percentage or through a binary score (e.g. completed vs. uncompleted)—may not be able to frame why and how the targeted strategic resources are expected to contribute to the organization's results. Such plans may not portray the performance driver(s) that the exploitation of these resources will affect, and the end-results it will influence in the long run.

Selective identification of performance drivers avoids confusion between means and ends. In fact, it engages planners in emphasizing the attributes on which project management must focus to contribute to the achievement of organization targets. For instance, launching a website might allow a department to communicate better, quicker and more efficiently than in the past with service users. Likewise, accomplishing a highway might allow a department to increase the carrying capacity of the transportation infrastructure, and therefore to reduce delays and

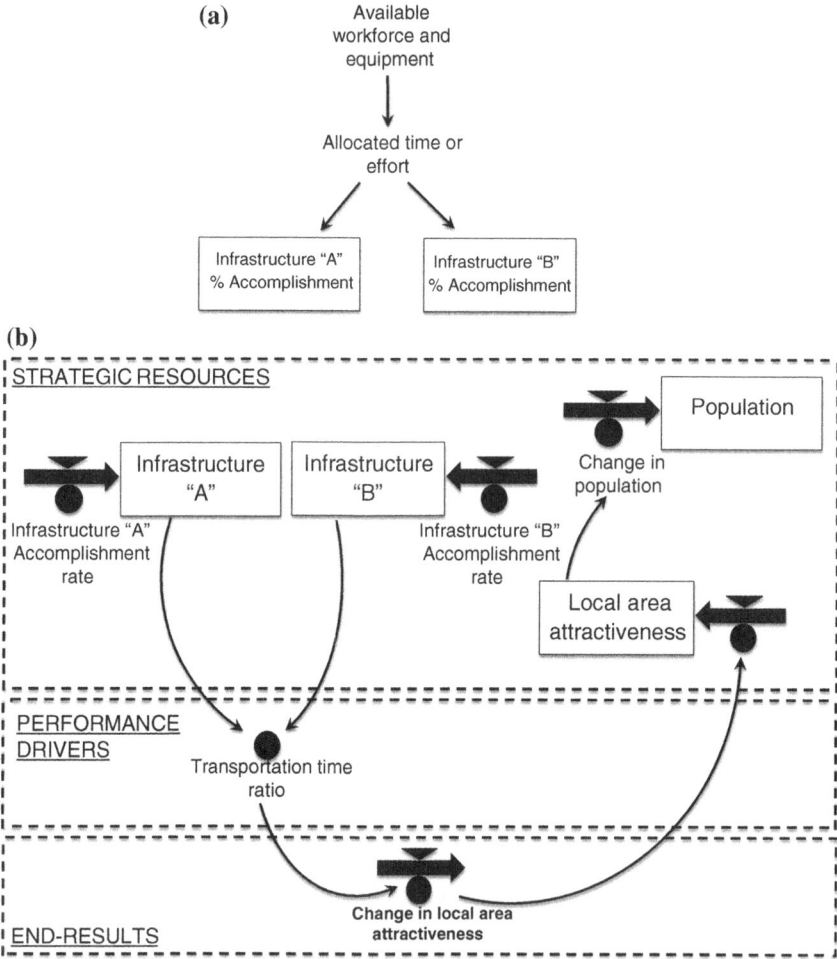

Fig. 3.7 a Example of a static and inconsistent of performance measures. **b** Example of a consistent DPM

accidents. This might lead to a higher attractiveness of the region, which would increase population.

By switching from static to dynamic performance management, one may shift the emphasis from only measuring the implementation of performed actions to measuring and managing their own effects over time, both in terms of outputs and outcomes (see Fig. 3.7a, b).

3.3.2 Performance Drivers and Performance Indexes

A second problem—related to the previous one—that may happen when DPM is introduced into organizations whose culture and practice are rooted into a static view of performance management is confusion between *performance drivers* and *performance indexes*.

Performance indexes are synthetic measures of the quality or state of a system. Likewise performance drivers, they are usually expressed as ratios. While drivers affect end-results or other performance drivers, indexes do not affect any specific performance measure. They rather gauge a synthetic expression of specific aspects regarding performance. Therefore, while performance drivers are relevant measures for performance *management*, performance indexes can be relevant for performance *measurement* only.[5]

When organizations try to "convert" static into dynamic balanced scorecards, they can be inclined to make inversions of causality by confusing performance indexes with performance drivers, and to connect them to the end-results. For instance, indexes like "Return on Investment"[6] or "Operating Leverage"[7] can be erroneously mapped as drivers of income, while they are an effect of it.

A few more detailed examples of common causation errors related to performance indexes and hints on how to reframe indexes into drivers will be now discussed.

3.3.2.1 Municipal Garbage Collection

Figure 3.8a, b show how the variables "Working time saturation" (the ratio between the working hours per week and a maximum working hours per week) and "Tons collected per .000 population" can be useful indexes to synthetically gauge efficiency and effectiveness in garbage collection by a Municipal agency. However, it would be a causation error considering them as drivers of the "Tons collected per week" (output end-result) and of the "Change in customer satisfaction" (outcome end-result). In fact, both indexes are affected by organizational performance, but they do not affect it.

As Fig. 3.8a shows, though the working time saturation is an efficiency index, the performance measure driving the end-result (tons of garbage collected per week) is "Tons collected per working hour". By multiplying such measure by the

---

[5]There are cases of measures that are at the same time performance indexes and can be modeled as performance drivers as well. This may happen when an index affect decision makers' perceptions and bias. An example is the "debts-to-total investments" ratio, which often affects the change in company solvency perceived by funders.

[6]Operating Leverage is the ratio between the total contribution margin (i.e. Sales revenues–Variable costs) and Operating income.

[7]ROI is the ratio between operating income and net investments.

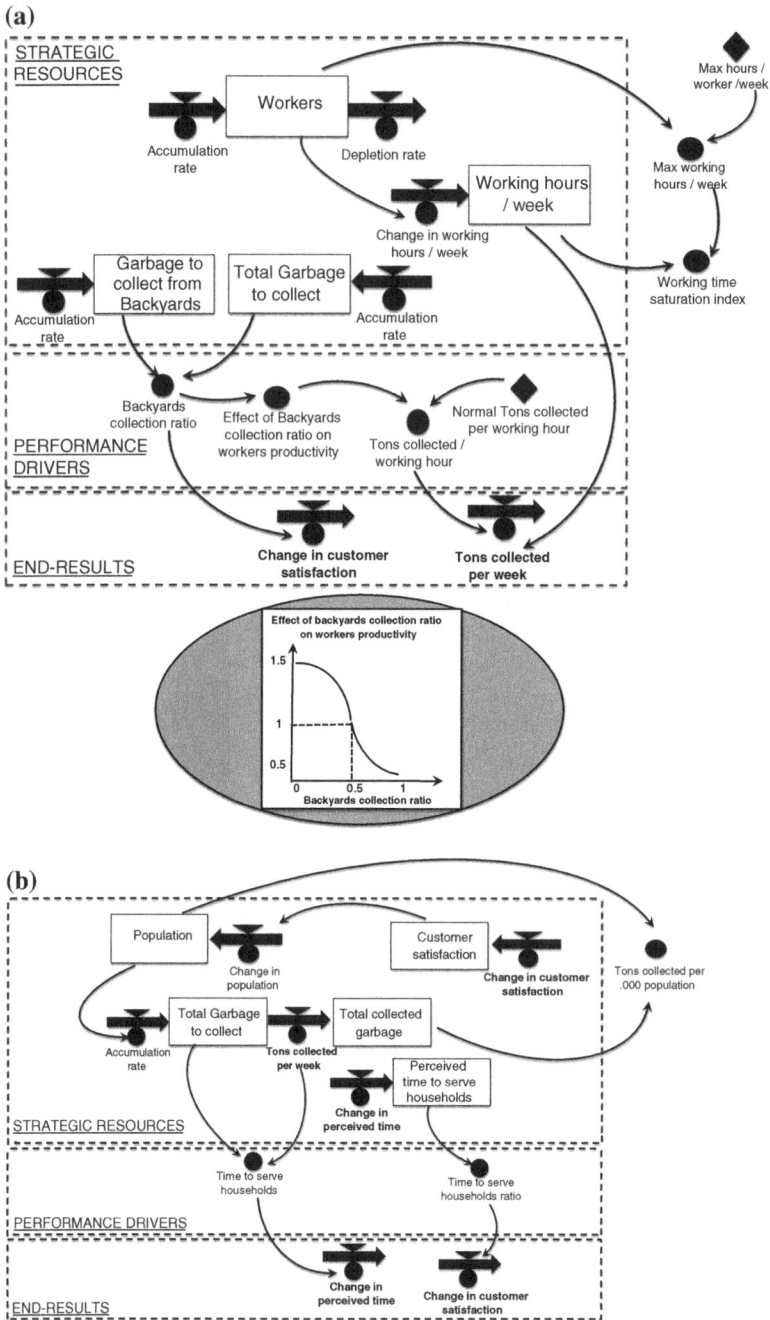

Fig. 3.8 a Identifying the performance drivers related to the working time saturation index in garbage collection workers' productivity modeling. **b** Identifying the performance drivers related to the "Tons collected per .000 population" index in garbage collection service provision modeling

working hours available per week (strategic resource), one may estimate the effect on workers' productivity associated with different alternative garbage collection methods. Though backyard garbage collection is less efficient than curbside pickup, administrators might consider it as a preferable option, since it might fit more than the other option with community's culture and identity.

A trade-off between efficiency (cost) and effectiveness (service) could be framed through DPM.

On the efficiency standpoint, an increase in the backyard garbage collection ratio (i.e. the percentage of garbage to collect from backyards on the total[8]) would decrease workers' productivity and vice versa. So the backyard collection ratio is a *second level* performance driver, affecting the "Tons collected per working hour" (*first level* driver) and—through it—the "Tons collected per week" (end-result). As shown in Fig. 3.8a, the effect of such *second level* on the *first level* driver can be modeled through a normalized graph function.

If observed from the effectiveness standpoint, an increase in the backyard garbage collection ratio would increase customer satisfaction (and vice versa). So, in respect to the change in customer satisfaction (outcome end-result), the backyard garbage collection ratio is a *first level* performance driver.

Figure 3.8b shows that, though the "Tons collected per .000 population" is an index providing a synthetic measure of service provision effectiveness, the performance measure driving the outcome end-result (change in customer satisfaction) is the "Time to serve households ratio". Such performance driver is calculated as the ratio between "Perceived time to serve households" and a benchmark (i.e., a reference, standard or desired time). As previously commented, the effect of such ratio on the end-result can be modeled through a normalized graph function.

The change in the perceived time to serve households (end-result) is in turn affected by the current "time to serve households", which is calculated as a ratio between the total garbage to collect and the tons collected per week.[9]

The figure also shows that a change in customer satisfaction will affect population dynamics, which will in turn affect the total garbage to collect, that will influence service levels.

From the discussed example it is possible to argue that the feedback loops profiling the structure and behavior of the investigated dynamic system could not be properly framed by simply linking performance indexes to end-results.

The next examples will give more insights on how to "convert" the performance indexes that information system's statistics may periodically deliver, into performance drivers that can be embodied by DPM charts and stock-and-flow simulation models.

---

[8]Both "Total garbage to collect" and "Garbage to collect from backyards" are here modeled as strategic assets, since they are information resources based on which the organization will make decisions aimed to combine efficiency with effectiveness.

[9]The change in the time to serve households equation is: (Time to serve households-Perceived time to serve households)/time to serve households observation time. This is a flow related to an information delay.

	Value	%
On time A/R collection	€ 20.000,00	13%
Delay > 0 wks <= 2 wks	€ 35.000,00	23%
Delay > 2 wks <= 4 wks	€ 45.000,00	30%
Delay > 4 wks <= 6 wks	€ 30.000,00	20%
Delay > 6 wks	€ 20.000,00	13%
	€ 150.000,00	100%

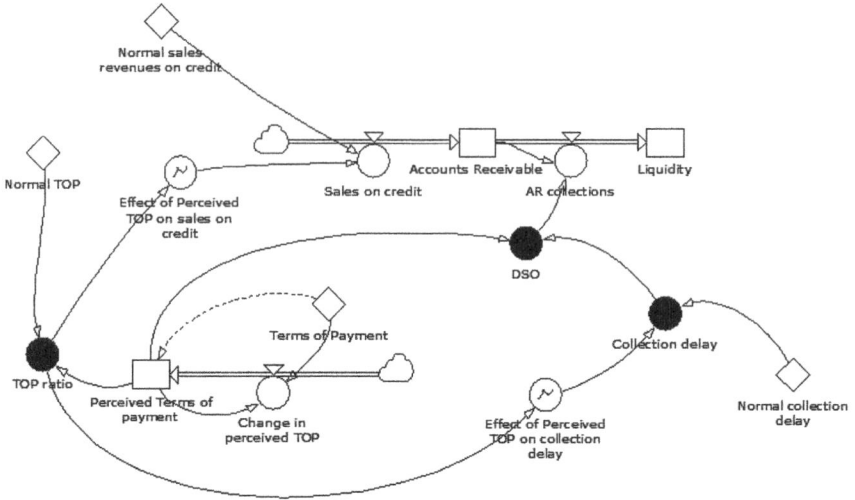

Fig. 3.9 Indexes from overdue accounts/receivable statistics and related performance drivers

3.3.2.2   Terms of Payment Policies

This example relates to a firm that is a supplier of other firms. With the intent to increase sales, it allows them longer terms of payment (TOP) than its competitors (normal TOP). However, while sales increase, the firm also experiences bad debts problems.

The table portrayed in the upper section of Fig. 3.9 shows that the firm is able to collect punctually only the 13 % of its sales on credit. The 30 % of credits is collected with a delay between 2 and 4 weeks. A stock-and-flow diagram, based on an instrumental view of DPM, can be sketched to portray how the terms of payment allowed by the firm to its customers drive sales orders and revenues on a side, and sales collections on another side. In such diagram, the values reported as percentages in the table are re-framed as performance drivers.

If the value of A/R at the beginning of the survey providing such indexes were Euro 150,000, then the days of sales outstanding (DSO) could be estimated as an average of the collection time in each of the three clusters, weighted by the value of outstanding A/R for each cluster.

This is a *financial performance driver*, since it affects A/R collection flows (see the stock-and-flow diagram in the lower section of Fig. 3.9).

This performance measure is still too aggregate to support trade-off analysis and policy making on the effects generated by aggressive terms of payment policies on the competitive (commercial) and the financial performance of the firm. In fact, the DSO is a sum between the reference terms of payment allowed by the firm[10] to its clients and a collection delay (*second level* financial performance driver) related to the clients' aptitude not to honor debts at the maturity date (CRIBIS 2014, pp. 142–157). Such collection delay can be estimated as an effect of a *third level* performance driver (in respect to A/R collections), i.e. the TOP ratio. This is calculated as a fraction between the perceived terms of payment allowed to customers and the normal TOP. The effect of such ratio on the collection delay can be estimated through a normalized graph function. The output of such variable will be the number of extra days—in respect to normal—customers would take to honor their overdue debts. In this case, the higher the TOP ratio is, the higher the clients' inclination to postpone their overdue debts' payment will be. In fact, by increasing allowed TOP, the firm will run a higher risk to sell its product to insolvent customers.

The TOP ratio is also a commercial performance driver, since it affects sales orders through another graph function that is a multiplier, similarly to the graph that was shown in Fig. 3.8a. Therefore, the sales on credit (end-result) will be equal to normal sales on credit multiplied by the effect of perceived TOP on sales on credit.

3.3.2.3 Product Portfolio Management Policies

This example relates to a firm that aims to expand its product maturity lifecycle. It pursues this goal by leveraging on marketing spend. Figure 3.10 shows the "marketing spend" parameter as a decision lever. It also depicts the perceived marketing spend as a strategic resource, since it is a measure of the company's marketing effort intensiveness perceived by the market in a given time. The marketing spend ratio (Average marketing spend/Normal marketing spend) is a *second level* performance driver, which affects (through a normalized function) the product time to reach late maturity.[11] This is a *first level* performance driver affecting the flow named "Mature products to late maturity" (end result). A reduction of such flow over time[12] (assuming a constant mature products stock) measures an improvement in product portfolio management.

---

[10]This is modeled in Fig. 3.9 through the variable named "Perceived Terms of payment". This is a delayed information variable through which the input terms of payment allowed by the firm are smoothed.

[11]The equation of "Time to reach late maturity" is: Normal time to reach late maturity x Effect of MKTG spend on time to reach late maturity.

[12]The equation of "Mature products to late maturity" is: Mature Products/Time to reach late maturity.

	Number of models to late maturity	%
> 0 yrs <= 2 yrs	6	30%
> 2 yrs <= 4 yrs	10	50%
> 4 yrs <= 6 yrs	4	20%
	20	100%

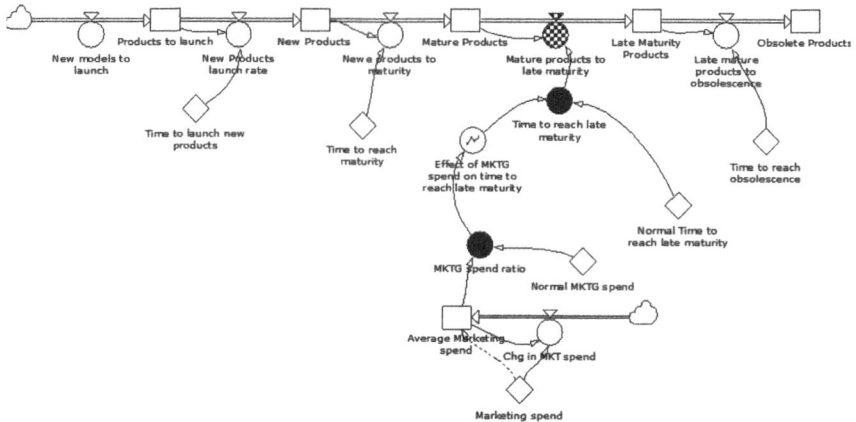

Fig. 3.10 Indexes from statistics on number of product models becoming late mature and related performance drivers

To calibrate the model functions measuring "Time to reach late maturity", the company information system may provide modelers with mature product models' lifetime indexes, like those depicted by the last column of the table in the upper section of Fig. 3.10. These indexes show that the 30 % of the stock of mature models reaches late maturity in less than 2 years; the 50 % in a time between 2 and 4 years; the remaining 20 % in a time between 4 and 6 years. If the number of mature models at the beginning of the survey leading to the measurement of such indexes was 20, the past time to reach late maturity could be estimated as an average between the average survival time of the models in each of the three clusters, weighted by the number of models going to late maturity for each cluster.

3.3.2.4 Telecom Industry

With the intent to support intellectual capital policies and to strengthen its customer base, a telecom service provider started a project aiming to "translate" its static BSC into an insight DMP model, based on an instrumental view.

Figure 3.11 frames the "competency index" as shown in the company BSC. The skill level and the knowledge level affect this measure. Training can influence both of them. The skill level gauges staff capability in reducing the time to: (1) fix billing

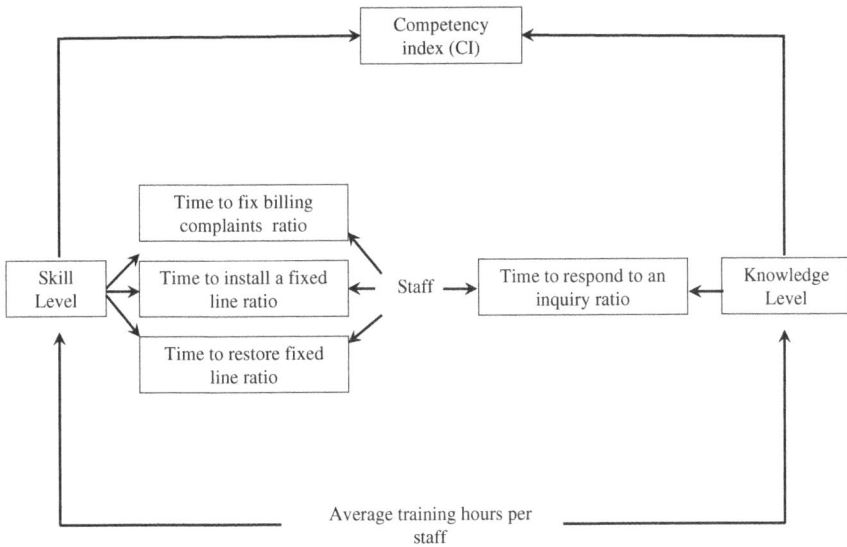

Fig. 3.11 Factors affecting the "competency index" in the telecom company's BSC

complains, (2) install a fixed line, and (3) restore a fixed line. The knowledge level gauges the staff capability to respond to customer inquiries and to effectively deal with them. Such delays contribute to determine the "customer service index".

In order to understand the impact of those indexes on the change in the customer base, we considered the competency index as a too aggregate measure to outline a set of performance drivers that could describe how and why customers could be gained or lost over time.

We started modeling the skill level as a normalized intangible asset, with values between zero and ten. As Fig. 3.12 shows, the skill level dynamics is governed by a balancing loop based on: (1) the desired skill level; (2) the gap between the current and the desired skill level, and (3) the percentage of the total training time allocated to skills versus knowledge improvement.

The skill level and the knowledge index are strategic resources affecting the main intellectual capital competitive performance drivers that influence customer acquisition and loss rates (end-results).

Figure 3.13 provides a comprehensive view of such factors impacting on customer base dynamics.

Concerning the customer acquisition rate, Fig. 3.13 shows the "combined product awareness and attractiveness ratio" as a *first level* performance driver. This is, in turn, affected by two *second level* performance drivers, i.e.: (1) the advertising spending ratio, and (2) the product awareness ratio. Other *third level* performance drivers affect the product awareness ratio, i.e.: (1) the "price ratio", (2) the "time to restore a fixed line ratio", and (3) the "time to install a fixed line ratio". The skill level and the knowledge index influence the last two drivers.

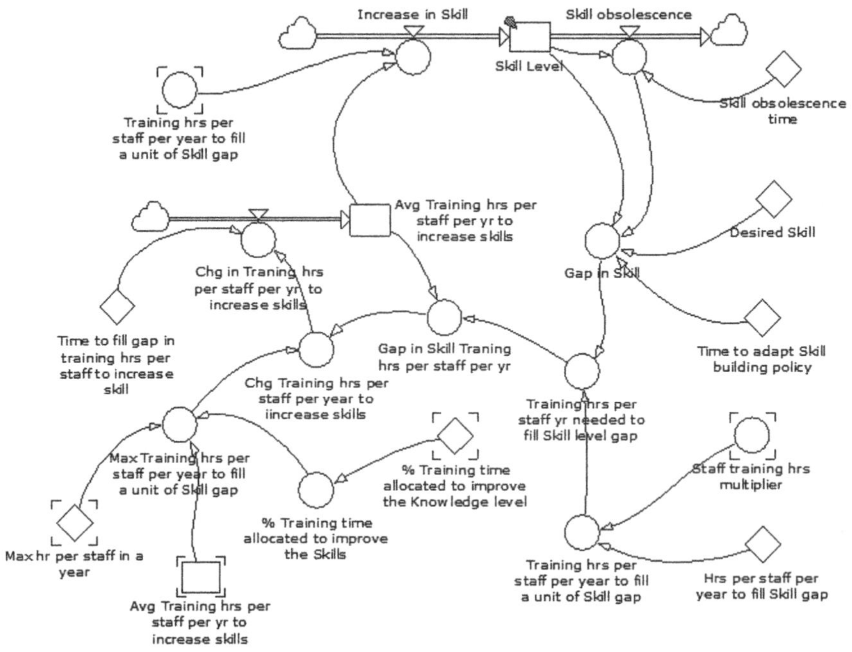

Fig. 3.12 Balancing loops affecting the skill level

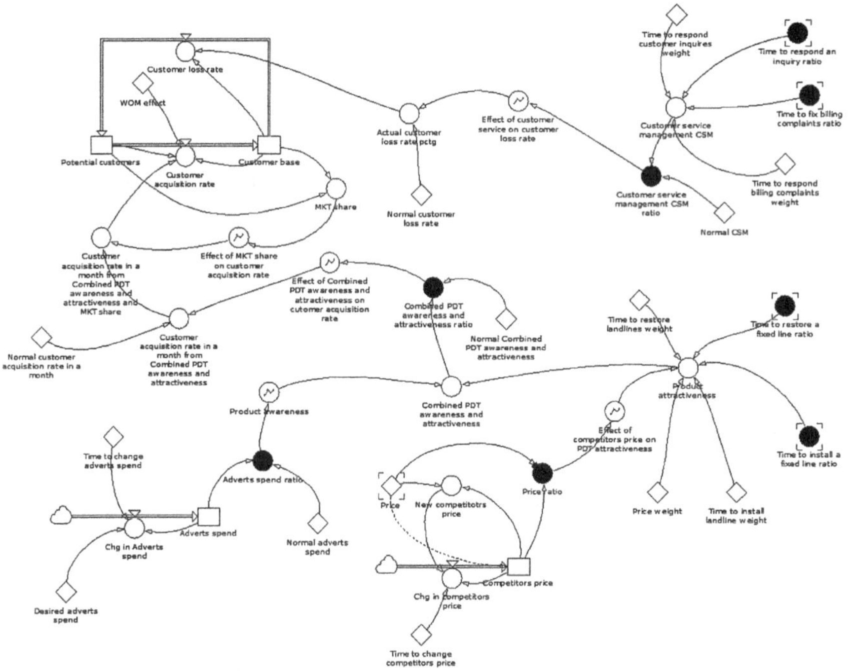

Fig. 3.13 Competitive performance drivers affecting the customer base

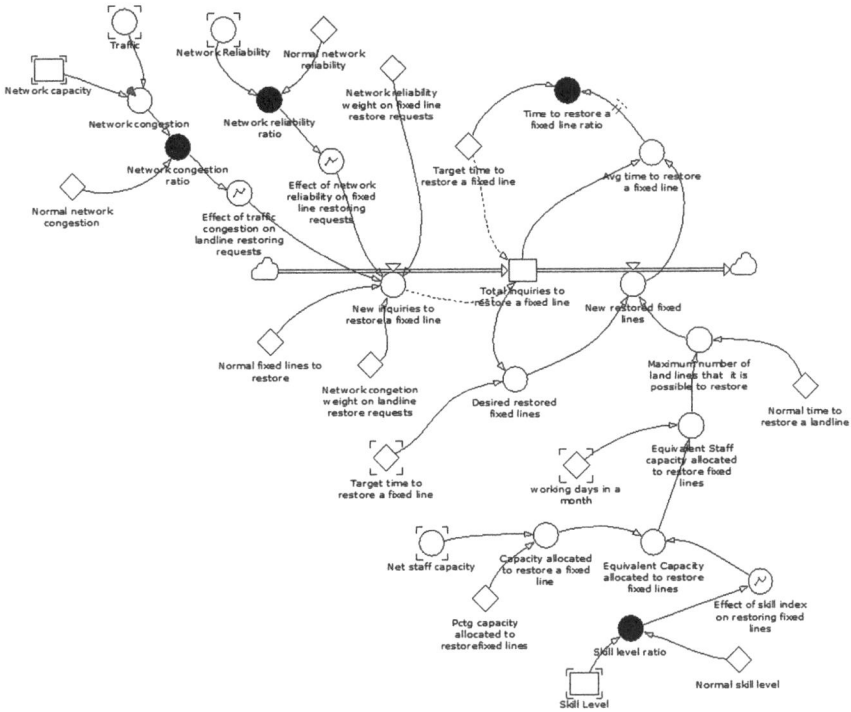

Fig. 3.14 Performance measures affecting the time to restore a fixed line

Concerning the customer loss rate, Fig. 3.13 shows that the "customer service management ratio" is a driver affecting the customer loss percentage. Other *second level* performance drivers affect such ratio, i.e.: the "time to respond an inquiry ratio" and the "time to fix billing complaints ratio".

For reasons of space we will now show how we modeled only one of the previously mentioned performance drivers related to the competency index. The approach we used has been the same for all of them.

The "time to restore a fixed line ratio" is a fraction between the average time and the target (or normal) time to restore a fixed line. The average time to restore a fixed line is the ratio between the total inquires to handle and the flow of restored fixed lines in a given time span (end-result). The main driver affecting this end-result is the skill level ratio: the more skilled staff is the less time fixing a landline will take.

Figure 3.14 also shows that another end-result (i.e., "new inquiries to restore a fixed line") is relevant to model the average time to restore a fixed line. In fact, it accumulates into the numerator of such measure: the more inquiries are received, the higher the time to fix them will be, other things being equal. Two more performance drivers affect such end-result, i.e.: the "network congestion ratio" and the "network reliability ratio". Improving them requires network investments and promptness in changing network capacity, respectively.

3.3.3   Applying the DPM Instrumental View to Competitive Performance Management on a Corporate Level

In this section two examples will be provided, in order to illustrate how an instrumental view of performance can be applied to competitive analysis in case-study class discussion.

"Systems Gear LTD" (Bianchi and Bivona 2002) produces and commercializes industrial machinery and equipment. The performance problem focused in the case is that—in spite of aggressive commercial efforts (price discounts, salesforce hiring, and product customization)—company sales orders have been declining. At the same time, the rising sales orders customization has generated production bottlenecks and increasing staff burnout.

The case discussion through an instrumental DPM approach, aimed to support planning and decision making to frame and solve the problem, helps class participants mapping the competitive drivers of sales orders and revenues and the strategic resources affecting them (Fig. 3.15). The analysis of key success factors in the business helps students identifying delivery delay, technology for production flexibility and invoicing errors as performance drivers which counteract the potential benefits on the sales order rate that aggressive commercial policies were expected to generate.

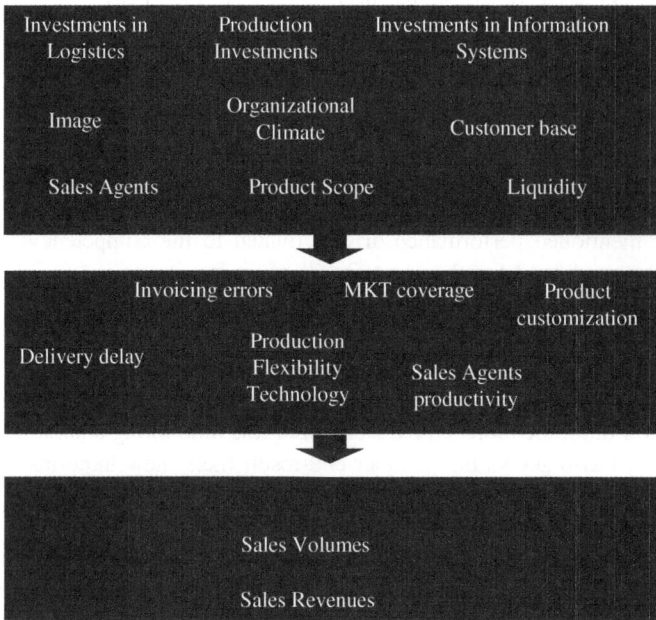

Fig. 3.15 DPM chart with an instrumental view: "Systems Gear" case-study

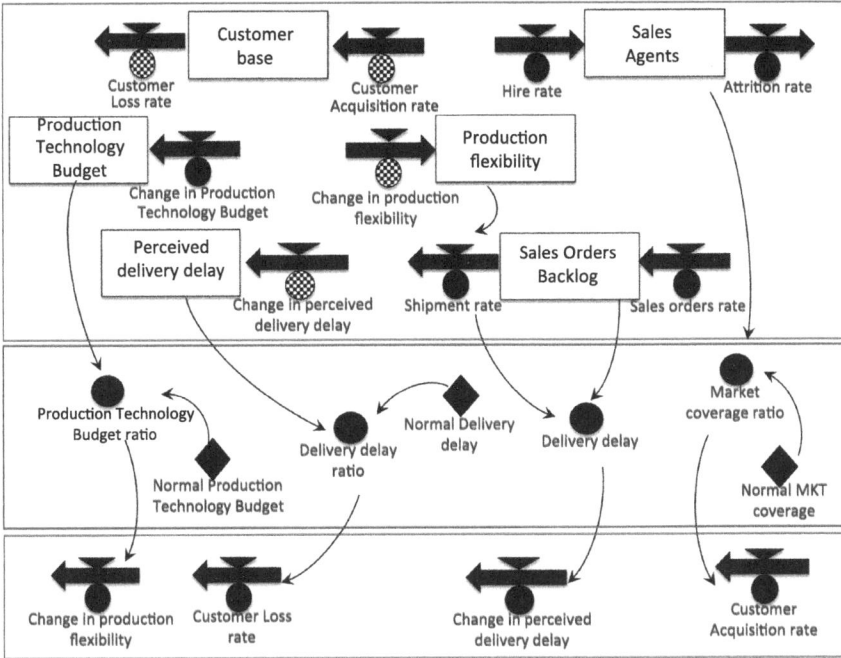

Fig. 3.16 Insight stock-and-flow DPM model based on an instrumental view: Systems Gear case-study

In a second step of the case analysis, a more detailed insight stock-and-flow DPM model—based on an instrumental view—is sketched and discussed (Fig. 3.16). This framework allows participants to outline the customer base as a strategic resource affecting the sales orders rate. They discover that a sustainable growth policy is not only related to the possibility to boost the acquisition of new customers (through commercial policies), but also to the need to drop the loss of existing customers. Therefore, two more end-result variables are added to the original map into the new insight stock-and-flow model.

In order to counteract the customer loss rate, a drop in the perceived delivery delay by customers must be fostered. Since a reduction in the sales orders rate is not an option, this requires that the product-shipping rate be increased. To this end, the company needs more production flexibility (leading to more capacity), which would allow the production department to deal with the rising customized product sales orders rate. Students then discuss how technology investments might increase production flexibility, which would increase the production and shipping rates. As Fig. 3.16 shows, the end-results are also depicted in the upper section of the DPM model, which frames strategic resources. To distinguish such flows from the others related to decisions (hiring/firing sales agents, accepting/fulfilling sales orders, investing in technology) aimed to affect directly strategic resources, end-result flows have been depicted by using a chessboard symbol.

Related to this model, class discussion leads participants to frame trade-offs in time and space. As a next step in this analysis, students are asked to convert the insight stock-and-flow conceptual model depicted in Fig. 3.16 into an insight simulation model.

The same method will be now used to frame another case, i.e.: the *Saturday Evening Post* (Hall 1976; Maciariello 1984, pp. 56–62; Wynne 1985).

The case describes the rise and fall of a magazine publishing firm. In this industry, competition is based on four main policy levers, i.e.:

(1)   The annual price of the magazine (i.e. the annual subscription price);
(2)   The price per insertion (i.e., the price paid by advertisers). The absolute value of the advertising rate is not the factor directly influencing the number of purchased advertising pages per issue. In fact, advertisers' decision depends not only on the advertising rate but also on the circulation of the magazine, i.e. on its total readers.
(3)   The annual promotional spending, aimed to raise both trial and regular subscriptions (i.e. circulation promotion);
(4)   The volume of the magazine (i.e. the pages per issue).

Since most publishers establish a ratio between the number of advertising pages and the total size of the magazine, the number of cumulative pages published in a year (and related costs) may significantly fluctuate over time.

Similarly to its three main competitors, the company was vertically integrated, with large capacity investments; this required a focus on a high sales volumes and revenues, coming from both a massive diffusion among subscribers and advertisers. This focus implied that—particularly in the '50s—the company undertook aggressive commercial policies, based on low subscription and advertising rates, as well as on massive promotional campaigns.

In spite of a progressive increase in the company magazine's total readers (which jumped from 3.25 millions in the 1940 to 6.30 millions in the 1960), the magazine's publication was discontinued in the 1960 because of the economic losses incurred by the firm. Such losses were caused by a rapid increase in production costs, which was not offset by a sufficient increase in the revenues.

As previously described, this case may be also used in class discussion by adopting an instrumental DPM approach, with the goal to enhance participants' capabilities to model the counterintuitive dynamic behavior of variables impacting on revenues and costs, and to enhance "intelligent" strategic and management control. Again, the key to conduct this analysis stands into a selective identification of performance drivers. The DPM chart depicted in Fig. 3.17 illustrates that—as a result of this first step in the case discussion—students identify different "layers" of end-results. This analysis allows one to measure such results not only through aggregated financial variables, but also through quantitative measures describing the organization's ability to build up strategic resources through management routines. These measures are: (1) the number of advertising pages sold per issue, (2) the number of article pages published per issue, (3) the change in the stock of

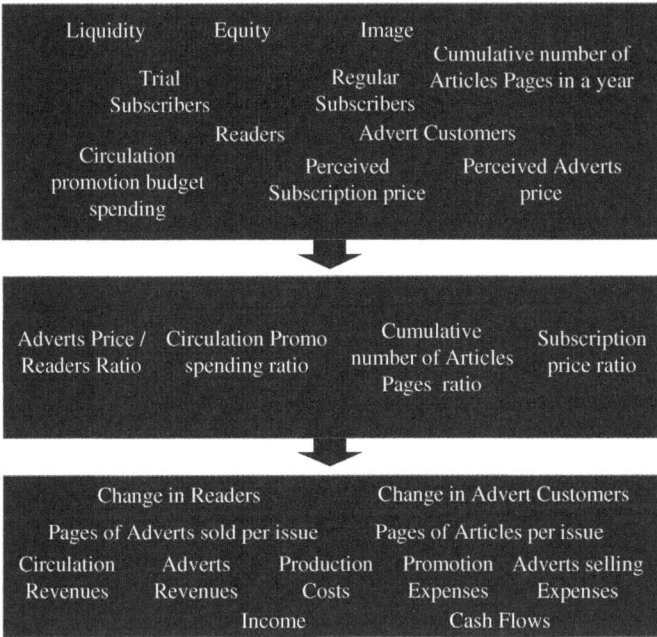

Fig. 3.17 DPM chart with an instrumental view: "Saturday Evening Post" case-study

readers (customer base), and (4) the change in the stock of advertisers (i.e. the number of customer firms from which advertising revenues come).

The set of strategic resources included in the upper section of the DPM chart embodies the factors affected by the mentioned end-results, i.e.: (1) the cumulative number of pages of articles in a year; (2) the stock of advertisers (customers); (3) the number of readers (trial and regular subscribers); (4) company liquidity and equity. The identification of the other strategic resources is made possible after the competitive performance drivers have been analyzed.

The chart illustrates the four most important drivers, i.e.: (1) the "advertising price/readers" ratio; (2) the "circulation promotion spending" ratio; (3) the "cumulative number of article pages" ratio; (4) the "subscription price" ratio.

The first driver affects the change in advertising customers and the number of advertising pages sold (and consequently the number of article pages) per issue. It also affects the advertising revenues. To increase such end-results, it is not enough to reduce the advertising rate. It is also important to increase the number of readers.

The other three drivers affect the change in readers, which in turn affects the circulation revenues.

In order to influence such drivers, the company policies should identify the strategic resources to build, deploy and combine for each of them. So, readers and the advertising price perceived on the market are the assets affecting the first driver. The circulation promotion budget spending is the resource affecting the second

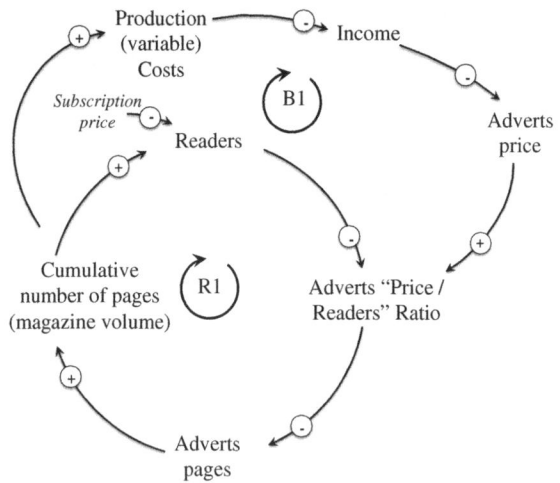

Fig. 3.18 Feedback loops associated with limits to commercial growth: "Saturday Evening Post" case-study

driver. The cumulative number of article pages and the total number of pages published in a year influence the third driver. The perceived subscription price affects the fourth driver.

An aggressive commercial policy may, on the one hand, increase readers, advertisers, revenues, and the total number of pages published in a year. However, the total number of pages published in a year also increases production costs. In order to reduce such costs, the company may decide to reduce the percentage of articles pages on the total published pages per issue. If such policy may lead to a higher income in the short run, it might generate stronger losses in the long run. In fact, publishing the magazine with higher percentage of advertising than competitors' magazines would reduce the company image (strategic resource). This would make commercial policies less effective. For instance, a drop in the regular subscribers might generate not only a reduction in circulation revenues, but also a higher "advertising price/readers" ratio. This would lead to a reduction in advertising customers and in the revenues.

Figure 3.18 illustrates the reinforcing loop (R1) related to the growth of readers and the magazine volume. It also illustrates the balancing loop (B1) providing limits to the commercial growth from the surge in production costs. This last loop implies that, to react to a lower income—due to higher production costs—the company erratically reacts by increasing the advertising price. This generates a reduction in the magazine volume and production costs.

In order to support policy making aimed to set sustainable development policies in the analyzed case, as a second step, students are asked to sketch a more detailed insight stock-and-flow DPM model, based on an instrumental view.

Figure 3.19 frames the loss of regular subscribers as an effect of the subscription price ratio (driver), which is affected by the perceived subscription price (strategic

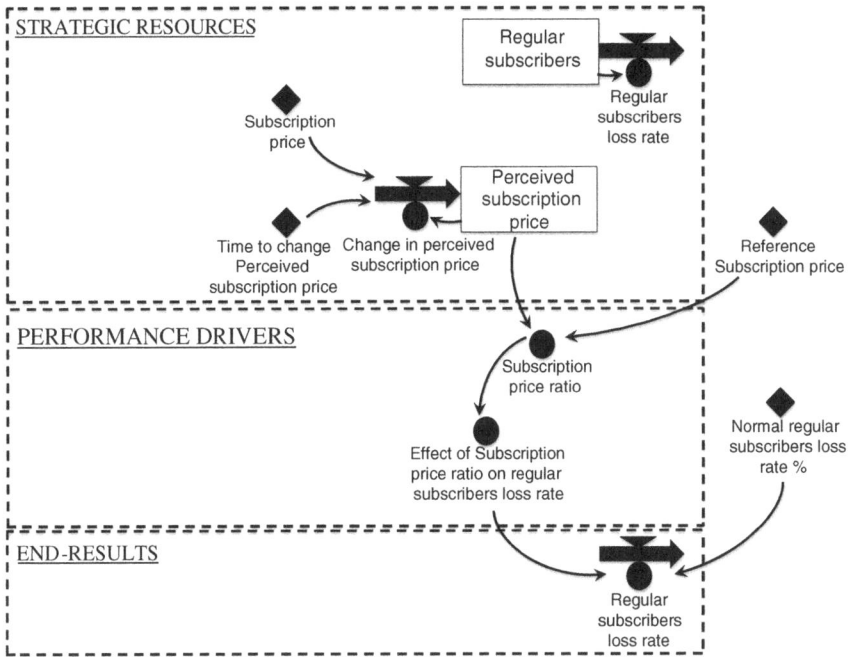

Fig. 3.19 Insight stock-and-flow DPM model based on an instrumental view: variables affecting the regular subscribers loss rate at "Saturday Evening Post"

resource).[13] The loss of regular subscribers is also mapped as an outflow from "Regular subscribers" (strategic resource).

Figure 3.20 illustrates the trial subscribers acquisition rate as an effect of the circulation promotion-spending ratio (driver), which is affected by the circulation promotion budget spending (strategic resource).[14] The trial subscribers acquisition rate is also mapped as an inflow to the "Trial subscribers" and as an outflow from the "Potential subscribers" (strategic resources).

Figure 3.21 shows the number of advertising pages sold per issue as an effect of the "advertising price/readers" ratio (driver), which is affected by the total number of readers (i.e., trial and regular subscribers) and the perceived advertising price.

---

[13]The equation defining the perceived subscription price is the same that is used in system dynamics to model information delays (Richardson and Pugh 1981, pp. 113–115), i.e.: Perceived subscription price = (Subscription price − Perceived subscription price)/Time to change perceived subscription price. Subscription price is the policy lever; time to change subscription price is a parameter measuring the speed at which subscribers are responsive to changes in price.

[14]Also the circulation promotion budget spending is a resource that should be modeled as an information delay.

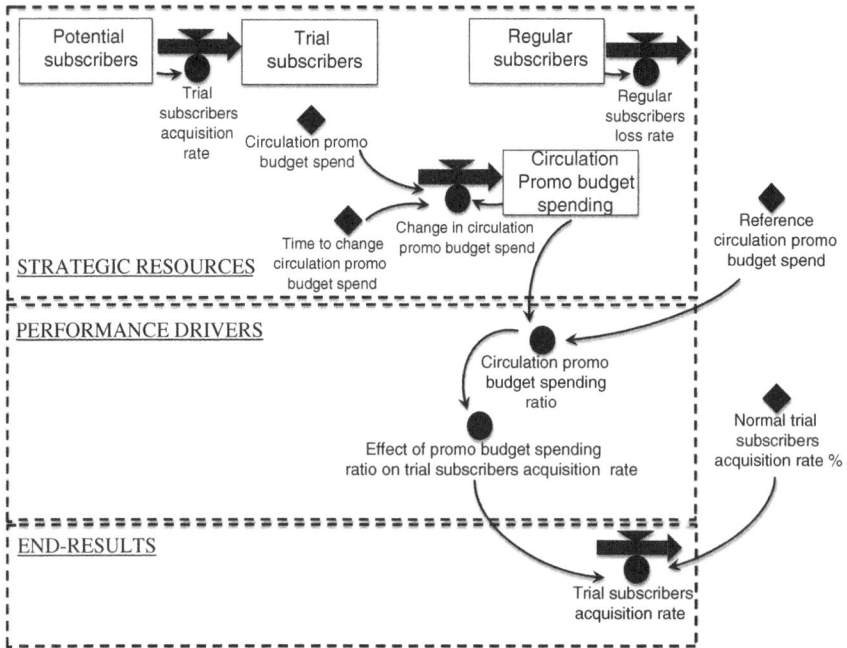

Fig. 3.20 Insight stock-and-flow DPM model based on an instrumental view: variables affecting the trial subscribers acquisition rate at "Saturday Evening Post"

Figure 3.22 shows the regular subscribers acquisition rate as an effect of the "cumulative number of pages" ratio (driver), which is affected by the cumulative number of pages in a year.[15] The regular subscribers acquisition rate is also mapped as an inflow into the "regular subscribers" and as an outflow from the "trial subscribers" (strategic resources).

DPM analysis supports learners in identifying the main cause of the company crisis, i.e. lack of control on the Advertising "price/readers" ratio, which is the driver depicted in Fig. 3.21. In fact, an increase in the magazine readers, caused by the company's commercial policies, generates an increase in advertising demand, even though the advertising rates are kept constant by the firm. The consequential surge in advertising pages—and, consequently, in the number of article pages—is the main cause of the experimented instability and crisis. By growing less rapidly than it did (i.e. by increasing advertising rates when the magazine circulation rises) the firm might have been able to contain the surge in the magazine volume and in production costs (see again loop "B1" in Fig. 3.18).

---

[15]The "published pages per issue" variable is calculated as follows: Pages of advertising sold per issue (1 + % Articles pages).

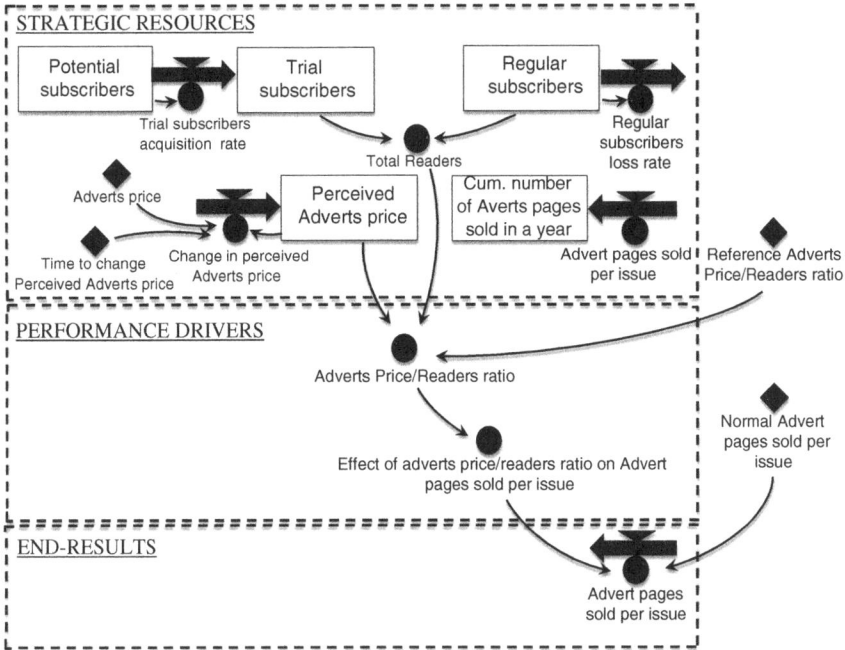

Fig. 3.21 Insight stock-and-flow DPM model based on an instrumental view: variables affecting the advertising pages sold per issue at "Saturday Evening Post"

This policy is framed in Fig. 3.23. The balancing loop "B2" illustrates that—by setting a desired Advertising "price/readers" ratio—an increase in the readers, leading to a decrease in the advertising price/readers ratio, would generate a negative gap between the desired and actual value of this ratio. Such a negative gap would suggest the firm to increase the advertising price. This would increase the Advertising "price/readers" ratio up to an—although transient—equilibrium state. The consequential reduction in the advertising pages might be beneficial for the firm.

Figure 3.24 summarizes through a stock-and-flow diagram the end-results affecting the dynamics of subscribers and the three main drivers impacting on them.

3.3.4   Cascading the DPM Instrumental View from a Corporate to a Departmental Level

The examples discussed in the previous section have shown how to apply an instrumental DPM chart to competitive performance management on a corporate level.

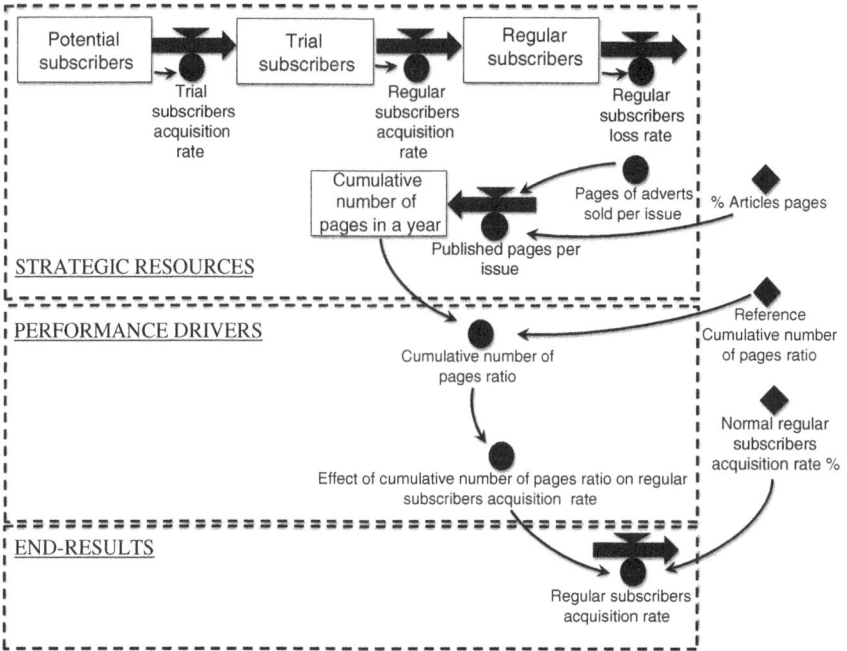

Fig. 3.22 Insight stock-and-flow DPM model based on an instrumental view: variables affecting the regular subscribers acquisition rate at "Saturday Evening Post"

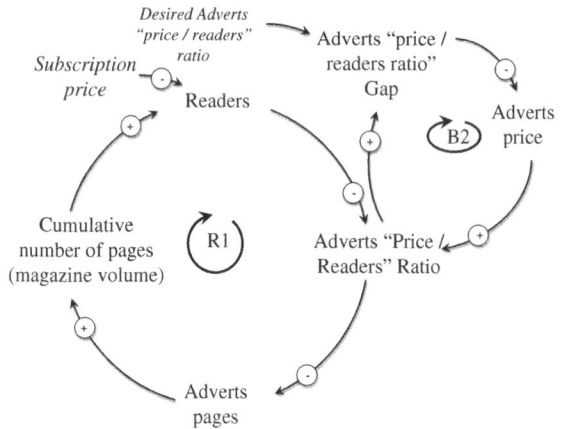

Fig. 3.23 A sustainable growth policy: leveraging on the advertising "price/readers" ratio at "Saturday Evening Post"

In this section we will analyze how to cascade it from a corporate (or divisional) to a departmental level.

Figure 3.25 shows a simplified version of the DPM chart that was previously illustrated in Fig. 3.5. It defines, on a corporate view, the product scope ratio and

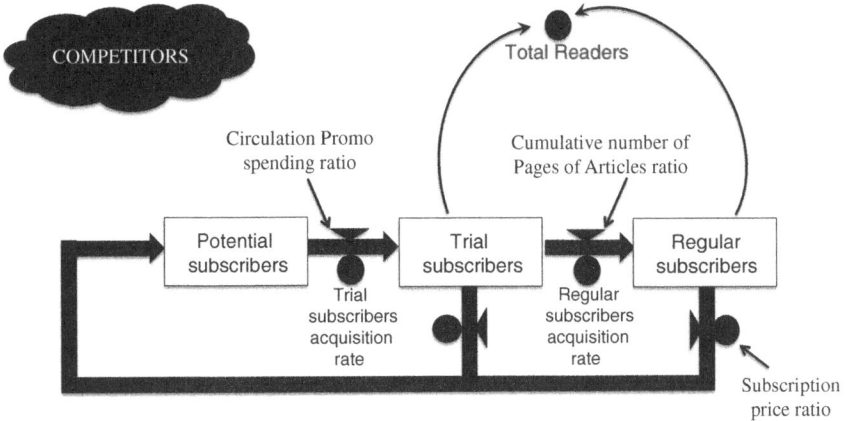

Fig. 3.24 The dynamics of subscribers and related performance drivers at "Saturday Evening Post"

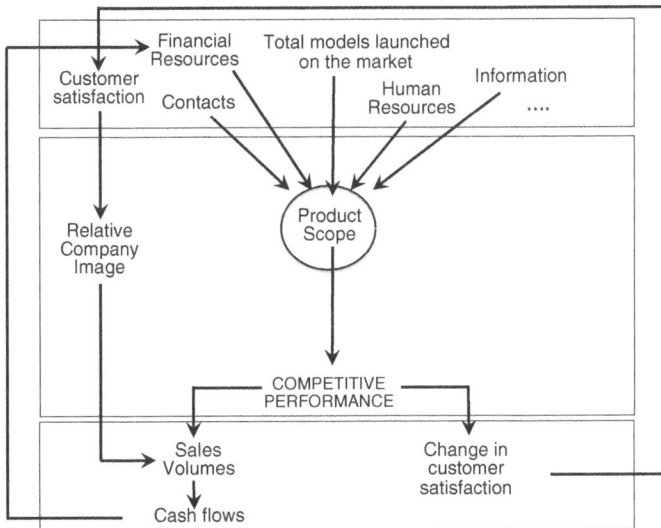

Fig. 3.25 A synthetic instrumental DPM chart on a corporate level

the relative company image as performance drivers. Several strategic resources affect the first driver.

Although such corporate DPM chart may have the merit to be synthetic, it might not be very helpful in understanding how management may build up and deploy such resources in order to affect the product scope. This issue refers to the problem we have discussed in the first chapter of this book about the need to coordinate strategic and management control. To this end, performance drivers should be

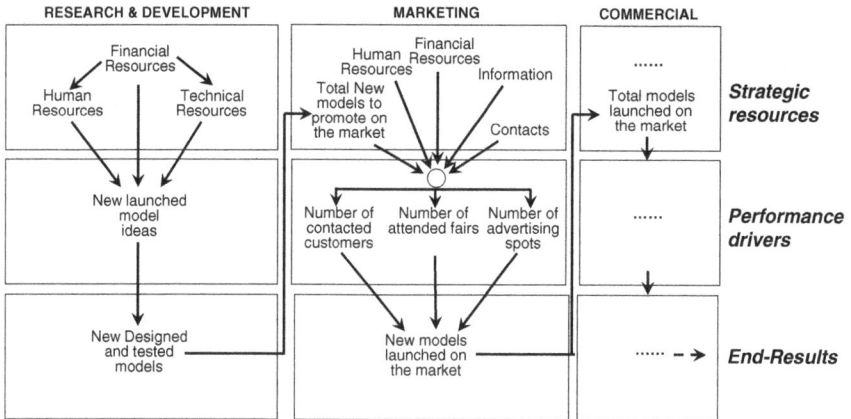

Fig. 3.26 Cascading the instrumental DPM chart to a departmental level

cascaded into the lower hierarchical levels of the organization, to understand how each department is expected to contribute to their achievement.

Figure 3.26 shows how a product scope measure, defined as a target on a corporate level, might be cascaded to three organization departments, i.e.: R&D, Marketing, and Commercial. An instrumental DPM view is replicated to the lower organizational levels.

The figure illustrates that, in order to improve the product scope, the commercial department might affect the company sales if the end-results delivered by the Marketing department will increase the stock of total models launched on the market (strategic resource). The Marketing department, in its turn, will be able to affect a target of new models launched on the market in a given time span (e.g. a 1-year budget) if: (1) the departmental drivers affecting such end-results will be identified and managed, and (2) the R&D department will allow Marketing to exploit a targeted stock of total new models to promote on the market.

This implies that R&D will set as a departmental end-result a given target of new designed and tested models in a given time span. It also implies that departmental performance drivers affecting such result and corresponding strategic resources will be identified.

Figure 3.27 clarifies how focusing end-results on a departmental level—as a consequence of cascading strategic corporate performance drivers—may allow one (in a third stage of the planning process) to better outline on a corporate level the performance drivers and end-results that will enable the firm to build up the strategic resources that cannot be purchased, such as in the case of the product models launched on the market.

A synthetic view of the analyzed feedback process is depicted in Fig. 3.28.

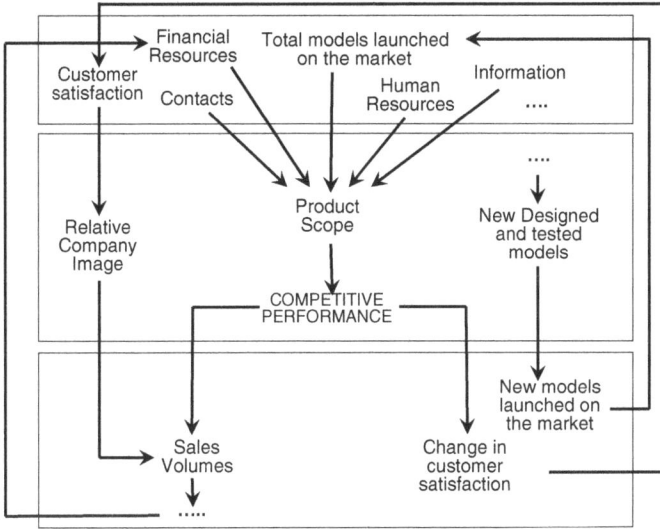

Fig. 3.27 Enhanced instrumental DPM chart on a corporate level

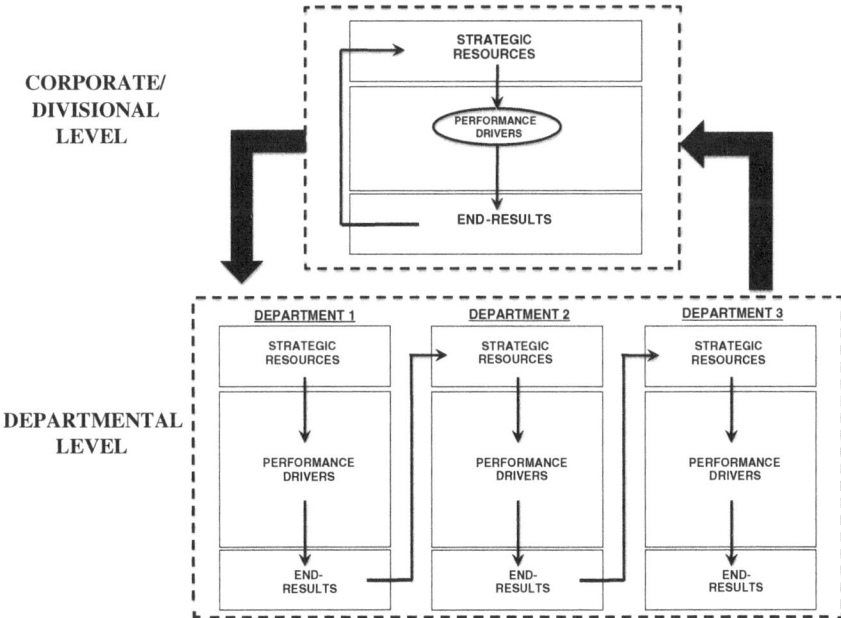

Fig. 3.28 A synthetic view of the instrumental view DPM process: from a corporate to a departmental level

3.3.5 Cascading the DPM Instrumental View
from a Corporate to a Departmental Level
(Continued): "Mosaicoon" Case-Study

Cascading a DPM instrumental view from a corporate to a departmental level may also allow an organization to start focusing the main "administrative products" that departments are expected to deliver in the fulfillment of the processes leading to the targeted end-results. This analysis prepares the field to the "objective" view of DPM, which will be discussed in the Sect. 3.4 of this chapter.

To provide another example of how this analysis can be performed, a case-study will be now discussed.

Mosaicoon is a small company whose core business is the development of innovative advertising "models" that combine the components of traditional advertising campaigns with those of interactive videos. This is referred as 'viral advertising' (Bianchi et al. 2015).

The company's product portfolio includes: (1) design, creation, distribution and tracking of viral videos; (2) development of web pages and institutional websites; (3) design of interactive tools for existing websites; (4) creation and promotion of interactive products; (5) viral diffusion of existing advertising videos on the web.

Figure 3.29 illustrates how liquidity, company image, quality of videos, and the capability of the firm to submit new commercial proposals, are the strategic resources that mostly affect both the flow of proposals accepted by customers and the rate of implemented projects. Such end-results pertain to the "Creative" area of the business. To pursue them, relevant performance drivers that this area should affect and constantly monitor are: (1) the relative price, and (2) the relative quality of videos.

The stock of total commercial proposals submitted by the firm is a precondition to improve accepted proposals and to implement projects. The commercial area affects the availability of such resource. To this end, the commercial area focuses on

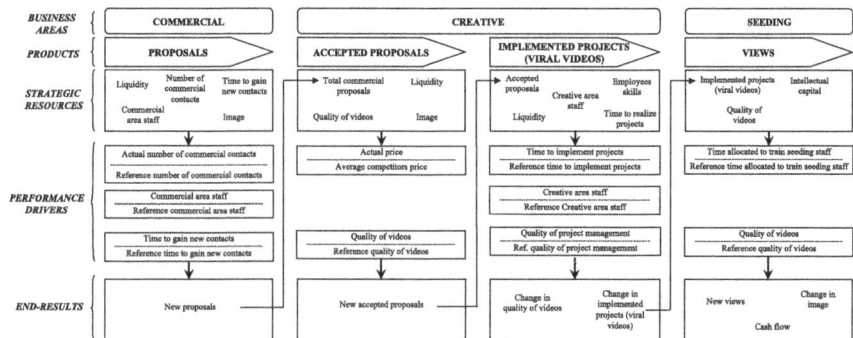

Fig. 3.29 Applying a DPM instrumental view on a departmental level: Mosaicoon case-study

the following performance drivers (ratios): commercial contacts, commercial staff, and time to gain new contacts.

Once a viral video has been made, it is ready to be broadcast on the web through a *seeding process*. Both the company and external partners share the videos on the web (e.g., on social networks and blogs), with the goal to deliver its customers a high quality service. This end-result is measured in terms of number of Internet views. While getting more views allows Mosaicoon to increase cash flows (which are proportional to the viral effect of the video), at the same time it improves its market image.

To affect such end-results, the "Seeding" area of the firm should influence the relative quality of videos (performance driver). This measure is affected by investing on intellectual capital (both staff development and selection of seeding partners).

3.3.6 Cascading the DPM Instrumental View from a Corporate to a Departmental Level (Continued): University Management

A second example, mostly related to public management, will be now discussed.

Figure 3.30 shows how an instrumental view of DPM may help a University in coordinating departments with central offices and faculties to sustain growth in research, teaching, and the placement of graduates.

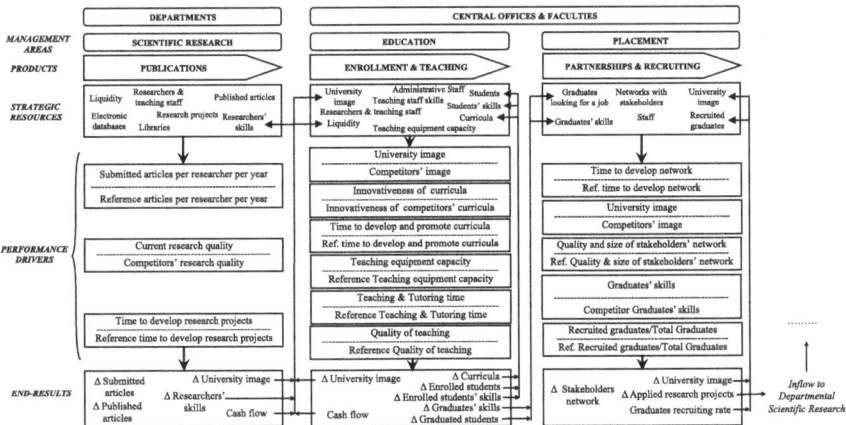

Fig. 3.30 Applying a DPM instrumental view on a departmental level: University management (adapted from: Cosenz and Bianchi 2013)

From the DPM chart it is possible to argue that the end-results generated through scientific publications by the University departments contribute to the improvement of three main strategic resources, i.e.: (1) University image; (2) researchers and teaching staff skills, and—if research is committed by funding stakeholders or its outcomes are transferred to them at a price—(3) liquidity.

Main performance drivers that may allow departments to pursue such end-results are related to the volume, quality and time through which research is carried out. Relevant strategic resources to affect such drivers are: staff (both in terms of size and skills), equipment, and liquidity.

University image, trainers' skills, and liquidity are strategic resources on which the outline, the promotion and delivery of quality teaching curricula can be carried out. Other strategic resources are: teaching equipment and administrative staff.

The better the quality and coordination of such resources is, the higher the number and level of skilled enrolled students and graduates will be (outcome end-result). Other output end-results related to the "Education" area are: (1) the number of enrolled students in a year, and (2) the change in the offered curricula. Main performance drivers affecting these end-results are related to: (1) quality of teaching, (2) time allocated to teaching and tutoring, (3) available equipment capacity (e.g. classrooms, labs), (4) time allocated to promote teaching curricula on the market, (5) innovativeness of curricula, and (6) University image.

The stock of the University graduates and their skill level are strategic resources on which the central offices may build to promote placement.

A first important end-result (an outcome) related to their activity is the graduate recruiting rate, which is an outflow from the stock of graduates looking for a job and an inflow into the stock of recruited graduates.

The main performance drivers affecting the volume of recruited graduates in a given time span are related to: (1) the graduates' skill level, (2) the size and scope of the university network with various stakeholders (e.g. enterprises), and (3) the University image.

A second outcome affected by the placement of graduates is the change in University image. This is affected by: (1) the percentage of recruited graduates of the total, (2) their skill level, and (3) the size and quality of the stakeholders' network.

A third end-result related to the "Placement" area is the change in the stakeholders network. This is an output measure that is affected by: (1) the time allocated by staff to develop the network, (2) the staff skill level, and (3) the University image.

A fourth outcome end-result measure through which "Placement" affects departmental scientific research is the flow of new applied research projects funded by networked stakeholders. Though this is not a result that can be explicitly linked to the graduates' placement promotion activity, it is strongly related to it. In fact, the previously commented measures (graduates recruiting rate, change in stakeholders network, change in university image) affect strategic resources that will, in turn, influence the probability that the University will agree with stakeholder partners to start new applied research projects funded by them or by third institutions.

3.3.7   Modeling Strategic Resources

The analysis developed in this section has identified five different kinds of strategic resources, i.e.:

1. **Physical resources** (e.g.: employees, machinery, inventory, points of sale). Such resources (or the services they provide) can be purchased. They are measured in *units*.
2. **Capacity resources** (e.g. manpower, machinery). They measure potential bottlenecks in the process of either transforming raw materials into end products or delivering a service. They are measured in units of production capacity *per time* (for instance: number of products/week).
3. **Information resources** (e.g.: orders backlog, expected demand, total working hours done). They are referred to either reports (including "coded" information) or organization decision makers' perceptions (implying "non-coded" information). Their unit of measure is the same as the reported or perceived variable.
4. **Resources generated by** *management (internal) routines* (e.g.: knowledge, business image, delivery delay perceived by customers, product quality, product portfolio, customer base). They cannot be purchased on the market. They are associated with: (a) perceptions that different stakeholders in the competitive and social system have about the organization. Such perceptions (e.g. regarding its reputation and solvency) can change by affecting the "first layer" end-results of the organization; (b) any other strategic asset the organization cannot purchase on the market, and can be built only inside on it (e.g. organizational climate, employees' burnout, morale, intellectual capital, capabilities, product quality, customer base, and product portfolio). If such resources refer to stakeholders' perceptions, their unit of measure is the same as the perceived variable. If they refer to assets that can be built up from only inside an organization, their unit of measure can be either dimensionless (as in the case of all intangibles) or in terms of units (such as for product portfolio and customer base).
5. **Financial resources** (e.g.: bank balances, equity, machinery value, inventory value, accounts receivable, accounts payable). They are reported in the organization's balance sheet (i.e. assets, liabilities, and equity). Their unit of measure is in currency terms.

Management implies that different types of strategic resources affect each other. Figure 3.31 shows an example of how to model physical and capacity resources. The rates at which the capacity stock changes over time are affected by the physical resource co-flows, multiplied by a "unit capacity" parameter.

Figure 3.32 shows that a physical resource depletion rate can be affected by:

- Information resources (e.g. raw materials inventory depletion rate depends on the sales orders backlog, which is an information resource);
- Capacity resources (e.g. raw materials inventory depletion rate is also affected by human and/or machinery production capacity);

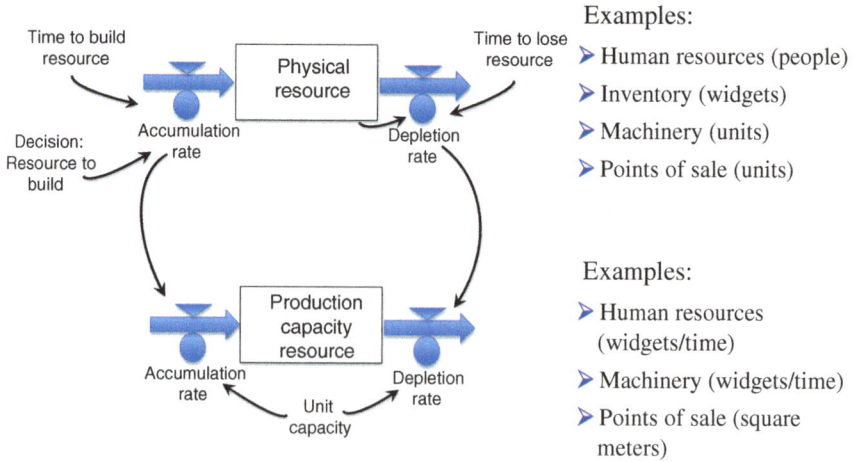

Examples:

➤ Human resources (people)
➤ Inventory (widgets)
➤ Machinery (units)
➤ Points of sale (units)

Examples:

➤ Human resources
  (widgets/time)
➤ Machinery (widgets/time)
➤ Points of sale (square
  meters)

Fig. 3.31 Relationships between physical and capacity resources

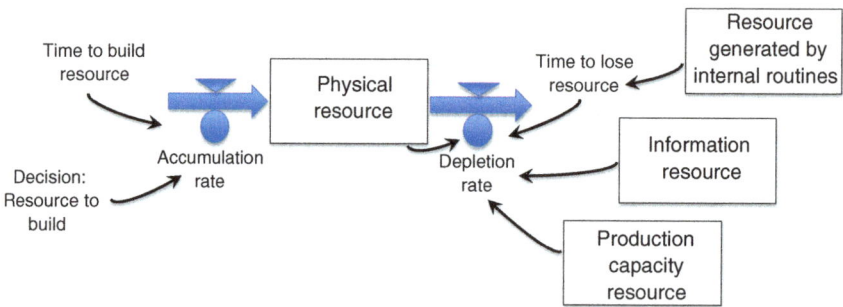

Fig. 3.32 Possible kinds of resources affecting the depletion rate physical resources

- Resources generated by *internal routines* (e.g. human resources attrition time can be affected by workers' motivation or morale, or by organizational climate).

The example in Fig. 3.33 shows how a physical resource can affect the outflow of a "coded" information resource. For instance, product-shipping rate from inventory (physical resource) can affect the depletion rate of the sales orders backlog ("coded" information resource). The figure also shows that the inflow of a "coded" information resource can affect the flow changing another "non-coded" information resource. For instance, the sales orders rate accumulating into the sales order backlog ("coded" information resource) may change the perceived sales orders rate.

Figure 3.34 shows an example of information resources affecting a resource generated by management (internal) routines. For example, product reliability perceived by customers (resource generated by internal routines) is affected by the

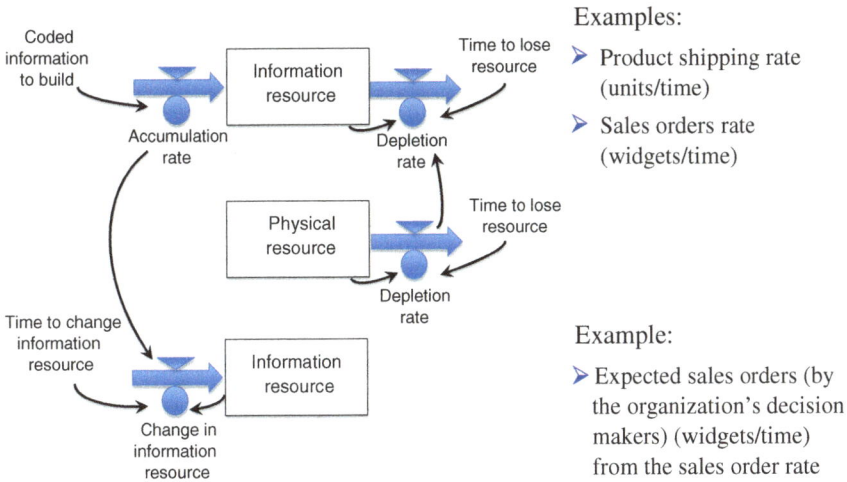

Fig. 3.33 Relationships between physical resources, "coded", and "non-coded" information resources

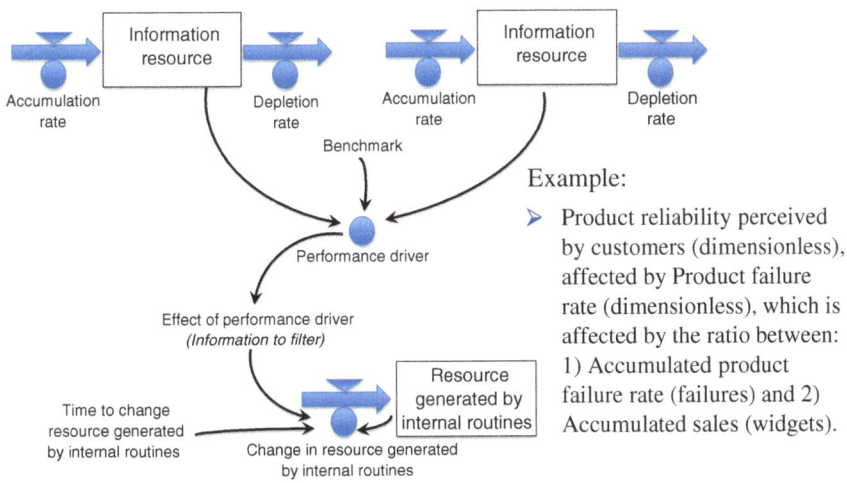

Fig. 3.34 Relationships between information resources and resources generated by internal routines

"product failure rate" (driver), which is the ratio between two information resources, i.e.: (1) "Accumulated product failure rate" (numerator) and a benchmark (denominator), and (2) Accumulated sales.

Related to the previous example, Fig. 3.35 shows another example framing possible relationships among resources generated by internal routines. The perceived product failure rate by customers ("non-coded" resource), may affect—through a

Example:

➢ Customer base (customers),
   whose retention rate is affected
   (through the time to lose
   customers) by the perceived
   product failure rate.

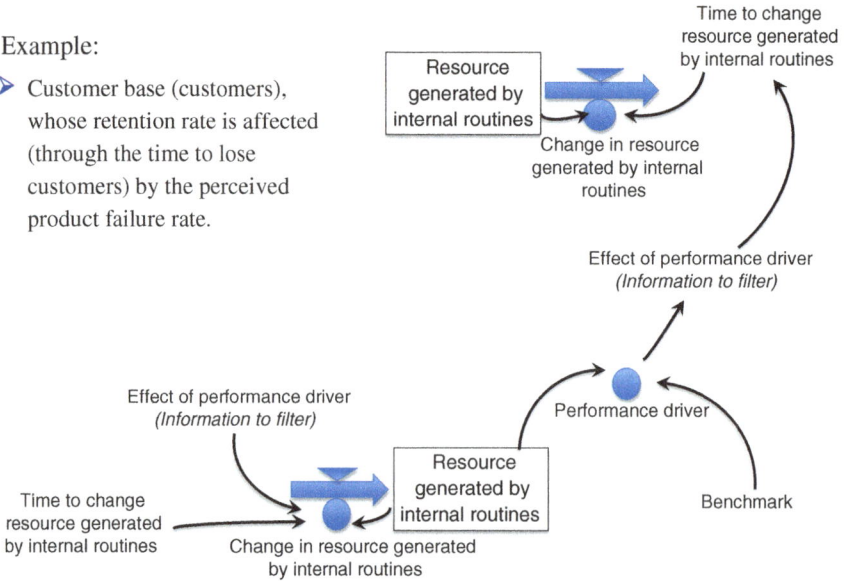

Fig. 3.35 Relationships among information resources

performance driver (ratio between such resource and a benchmark)—the average time to lose customers, which is the denominator of the customer loss rate, i.e. an outflow of the customer base ("coded" resource).

The structure portrayed in Fig. 3.36 shows how physical resources affect financial resources. This is possible through co-flows. For instance, equity (financial resource) is affected by income (inflow to such resource), which depends on the revenues from quantities of shipped physical resources (outflows from end-product inventory) and costs from purchased physical resources (inflows to raw material inventory). We may use the same structure to frame how: (1) liquidity (strategic resource) is affected by cash flows; (2) accounts receivable (strategic resource) is affected by the revenues from widgets sold on credit, and (3) accounts payable (strategic resource) is affected by the costs of raw materials purchased on credit.

Figure 3.37 shows how financial resources can affect physical resources. For example, the machinery acquisition is affected by available liquidity, as well as unit costs and the time to build up such resource.

Figure 3.38 frames how a financial resource can affect another financial resource, which in turn affects a resource generated by internal routines, that will affect a physical resource. For example, available liquidity affects the average intellectual capital investment (financial resource). This affects employees motivation (resource generated by internal routines), which affects the stock of personnel (physical resource) through the attrition time.

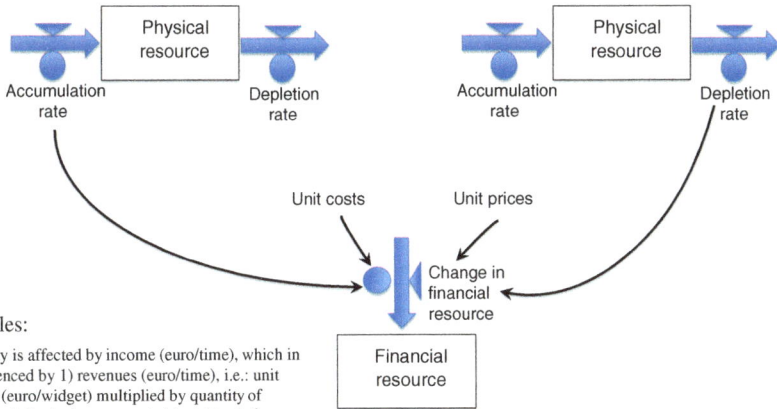

Examples:

➢ Equity is affected by income (euro/time), which in influenced by 1) revenues (euro/time), i.e.: unit price (euro/widget) multiplied by quantity of shipped physical resources (widgets/time). Same for the costs from purchased physical resources.

➢ Same for: 1) Liquidity, (affected by cash flows); 2) Accounts Receivable (affected by the revenues from widgets sold on credit in a time flow); 3) Accounts Payable (affected by the costs of raw materials purchased on credit in time flow).

Fig. 3.36 Relationships between physical and financial resources

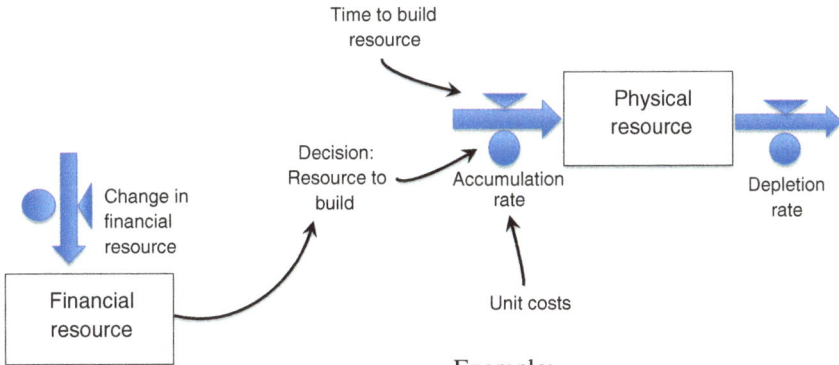

Example:

➢ Liquidity (Euro) affects the acquisition of machinery, based on its unit cost and the time to build such resource.

Fig. 3.37 Relationships between financial and physical resources

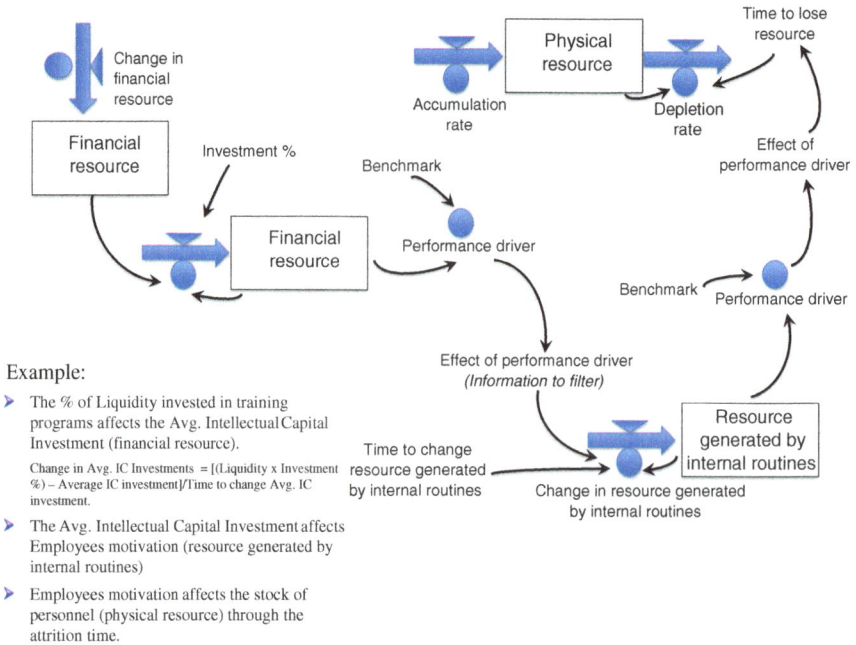

Example:

➤ The % of Liquidity invested in training
  programs affects the Avg. Intellectual Capital
  Investment (financial resource).

  Change in Avg. IC Investments = [(Liquidity x Investment
  %) – Average IC investment]/Time to change Avg. IC
  investment.

➤ The Avg. Intellectual Capital Investment affects
  Employees motivation (resource generated by
  internal routines)

➤ Employees motivation affects the stock of
  personnel (physical resource) through the
  attrition time.

Fig. 3.38 Relationships among financial resources and between them and: (1) resources generated by internal routines, and (2) physical resources

3.3.8 Implicit Modeling of Performance Drivers in System Dynamics

To conclude the analysis of how to model performance drivers through a instrumental view of DPM, a few more thoughts can be useful in relation to the possibility to identify implicit performance drivers in SD models.

Figure 3.39a provides an example.

This stock-and-flow model shows the sales orders rate (end-result) as an effect of the stock of sales agents (strategic resource) multiplied by and the average sales orders per agent (parameter).

This simplified model only apparently does not embody a performance driver affecting sales orders through sales agents.

It, rather, underlies a number of implicit hypotheses, i.e.:

1. The fixed parameter measuring sales agents' unit productivity ("Avg. Sales orders per agent") is related to an implicit standard performance driver. If, for example, the parameter equals to 100 «widgets/agent/week», and if the initial stock is 10 «agents», then—at the beginning of the simulation—the driver (i.e. the "sales agents ratio") will be equal to the ratio between the company's and "normal sales agents", which equals 10 salespeople.

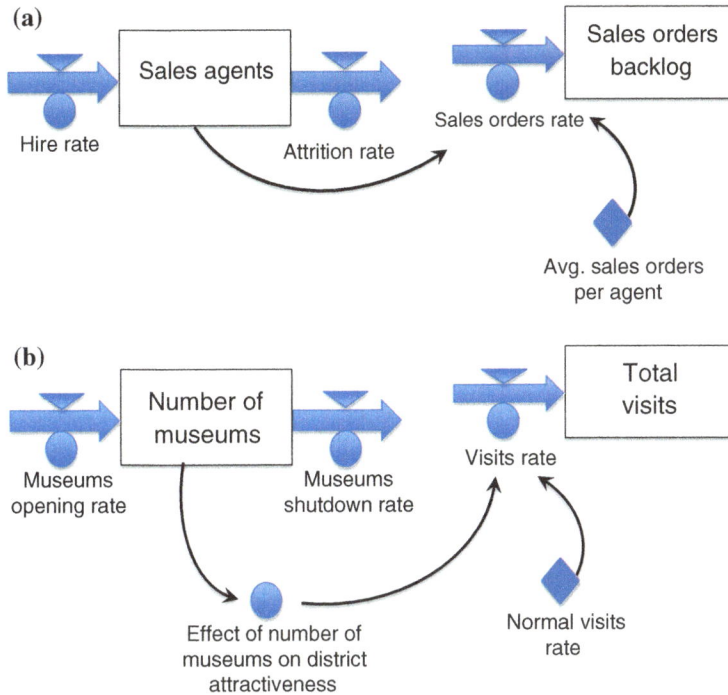

Fig. 3.39 **a** Implicit identification of performance driver in modeling how sales agents affect the sales orders rate. **b** Implicit performance standard identification in modeling the effect of the stock of museums on the visits rate

2. A linear relation exists between sales agents and sales orders. Therefore, the sales orders multiplier measuring the effect of the "sales agents ratio" on sales orders is proportional to the "Avg. Sales orders per agent" parameter.

This modeling approach has, undoubtedly, the merit to provide a simple feedback structure in respect to the alternative one that would make the performance driver explicit. To develop a SD model supporting DPM, this approach can be recommended if: (1) non-linear relationships (e.g. due to market saturation and competition) regarding the effect of salesforce on sales orders are not significant, and/or (2) the sales agents stock is not subject to significant changes in simulation time horizon.

If these conditions do not occur, such modeling approach implies a risk of oversimplification, particularly when a DPM approach is adopted. In fact, it both lacks emphasizing the process through which sales agents would allow the firm to receive sales orders, and it also implies that a linear relationship between sales agents and sales orders exists.

On the contrary, there might be diminishing returns in such relationship. For instance, an alternative performance driver—such as the "market coverage ratio"

(Agents/Potential customers)—that might improve the capability of such model to explain how hiring salespeople can affect sales orders.

Also, an additional performance driver affecting sales orders might be identified; e.g. "market share" (customer base/total customers), which might affect sales agents productivity as a multiplier.

A similar example (Fig. 3.39b) refers to the effect of the number of museums in a region (strategic resource) on the visits rate (end-result).

Defining a graph function (such as the "Effect of number of museums on district attractiveness") that is multiplied by the "normal visits rate" (fixed parameter) to affect the flow of visits implies the following:

1. The graph function implicitly embodies a performance driver (i.e. ratio between the number of museums in the district and the "normal number of museum"). In this simplified model, the total number of museums in the district directly affects the attractiveness multiplier. Therefore, the "normal number of museums" will be the value in the "x" axis of the graph function to which a value of 1 will correspond in the 'y' axis. For instance, if the value of 10 museums corresponds to a multiplier of 1, this will be the performance standard for this strategic resource.
2. If the parameter equals, for instance, 100 "visitors/museum/week", this will be the volume of visits (end-result) that is expected if the number of museums (strategic resource) equals to the normal (i.e. the standard of 10 "museums").

The comments to this feedback structure are very similar to those related to the previous one. Therefore, the structure has the advantage to be simple—and therefore—easy to frame, particularly if it is embodied into a much larger model. However, also this structure may imply the risk of hiding an important performance standard, and the process through which the end-results can be affected.

3.3.9 Comparing the Instrumental View of DPM to the Dynamic Resource-Based View

The literature and practice on SD applied to strategic management has developed a complementary approach to the described "instrumental" view of DPM, implying the study of the dynamic conditions leading an organization to build up and defend its competitive advantage over time. Such an approach has been defined as Dynamic Resource-based View (DRBV).

Previous studies adopting a DRBV of organizations have confirmed that the management of strategic resources, and more specifically the maintenance of an appropriate balance between such assets, is the key to sustainable development (Morecroft 2007; Warren 2008).

The emerging models all focus on the building up and decline of core assets, including workers, equipment, population, workload, perceived service quality, and

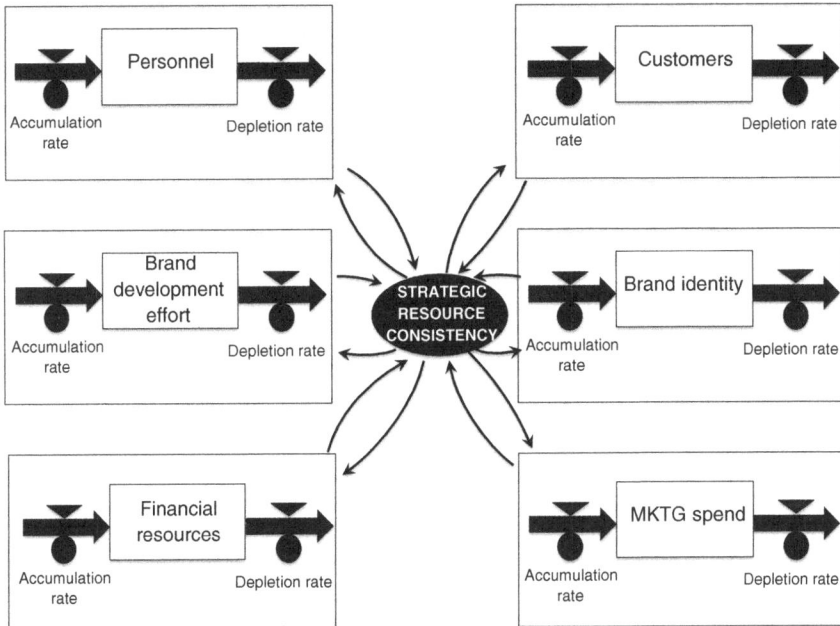

Fig. 3.40 A dynamic resource-based view

financial resources. Each of the strategic resources can, to some extent, be controlled in isolation of the others; however, where there is not balanced growth or coherence in the assets, then organizations will likely be unable to grow to achieve their own potential, or might grow in a non-sustainable way. The two common features in strategic resource management, as shown in Fig. 3.40, are the requirement for consistency between strategic assets and the need to actively manage each strategic asset to maintain balance.

A synthesis of both approaches is portrayed in Fig. 3.41. The figure emphasizes how the primary focus on resource consistency and on competitive advantage related to the DRBV may enhance an analysis of how performance drivers and related strategic resources can be modeled through the instrumental view of DPM.

On the other hand, the primary focus of DPM is not on the analysis of the conditions ensuring a firm to build up and defend its competitive advantage. It is, rather, on enhancing P&C systems through a "dynamic" and "intelligent" reporting of performance measures, and on providing decision makers with suitable keys to frame how and why such measures may change over time, as a function of adopted policies and of external factors.

Given the focus of DPM on performance measures, the conceptual starting point of this modeling approach is provided by the identification of the end-results and, as previously shown, of the related performance drivers, based on which strategic resources will be identified. The focus of this approach on performance measurement/management (at both the 'end-results' and 'drivers' levels) may allow

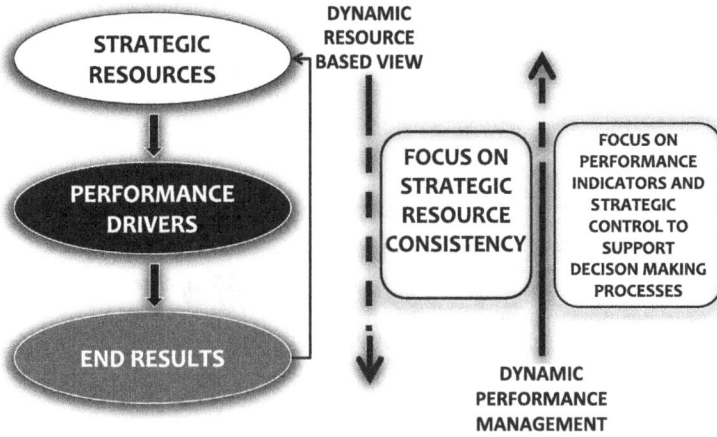

Fig. 3.41 A synthesis of a DRBV and the instrumental view of DPM as complementary approaches to manage organizational growth in a sustainable development perspective

controllers to enhance formal planning, reporting, and accountability on such measures. It may also allow them to improve the design of performance management systems.

Based on these remarks, the described logical sequence "end results → performance drivers → strategic resources" related to an "instrumental" DPM analysis, can be defined as a *P&C perspective*. In fact, it primarily aims to feed planning, reporting and performance evaluation.

On the other hand, if one considers the *decision-making perspective*, related to the implementation of plans, and to the undertaken actions to generate results that will be framed into performance reports, then the logical sequence to implement an instrumental DPM view is the opposite. In fact, decisions will build up, deploy and combine strategic resources, to affect drivers, which will influence the end-results, which will in turn affect those strategic assets that cannot be purchased through market negotiations. Both perspectives complement each other (Fig. 3.42).

Fig. 3.42 Sketching an instrumental DPM view: a P&C and a decision-making logical sequence

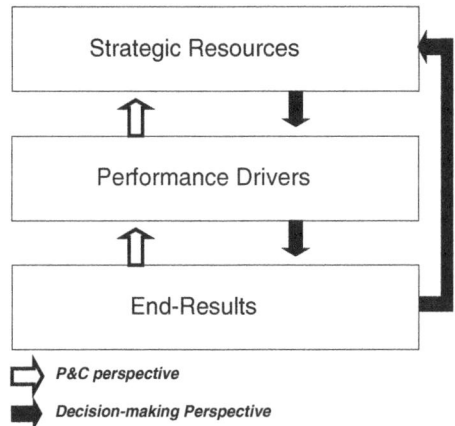

3.4 The Objective View of Performance

In the previous section it was observed that a corporate or divisional level of analysis through the instrumental view may allow one to define aggregate relationships affecting organizational performance. It has been also remarked that, by cascading corporate performance drivers on a departmental level, a closer analysis can be carried out.

Applying the instrumental view could be considered as an exhaustive way to implement DPM if:

- A simple organization structure (both in terms of responsibility levels and disaggregation of roles) does not require any further detail. This is a typical context that can be found in small businesses.
- The impact of back-office and support activities on performance drivers is marginal,
- The organizational culture towards a "dynamic" view of performance management is weak, and
- The client-organization cannot rely on sufficient financial resources and professional skills to sustain a pervasive DPM project.

The main goal of designing and implementing only an instrumental view is not to integrate DPM into formal P&C and evaluation systems. It is, rather, to support strategic planning by framing (also through insight SD simulation) the aggregate causal relationships among factors portraying organizational performance over time.

By integrating DPM into formal P&C and evaluation systems, one may better identify: (1) the performance drivers which mostly affect the end-results, (2) how decision-makers can concretely affect them, and (3) how decision-makers can be made accountable on them.

However, to pursue these goals, it is necessary to extend the instrumental view and to link it to two other views of performance management. Such views refer to:

- *How* the organization is expected to meet the performance drivers and the end-results identified through the instrumental view (the *objective* view of performance), and
- *Who*, in an organization, is responsible for the fulfillment of the activities that will allow one to deliver the associated "products" (the *subjective* view of performance).

This section will illustrate the objective view of DPM. Such view adopts a combined external and internal perspective (in respect to the organization). In fact, first of all it requires the identification of the *clients/users* or, more broadly speaking, the stakeholders with whom it interacts. It also requires that the *products* or *services* and social benefits it delivers be outlined. The "clients-products" of an organization should be identified under a competitive and social profile.[16]

Under the *competitive profile*, such analysis must focus on the definition of both the groups of clients or users whose needs are satisfied by an organization, and the products or services delivered to them.

Under the *social profile*, such analysis must focus on the definition of different groups of stakeholders towards an organization, and of the social benefits provided to them.

Identifying a "clients-products" system allows one to better focus the output and outcome end-results related to the instrumental view. The concept of "product" should not be restricted exclusively to the final result of manufacturing or commercial activities. Also, when an organization delivers a physical product, the service outcome can often be extended to a wider *product system*, i.e., an "offer package" that is consistent with key success factors for excelling in competition to serve clients and other stakeholders.

That is the case with *pre* and *post*-sale assistance. The concept of "product" also relates to the support of clients in inventory management, invoicing, and more broadly in collection-cycle management. In all these cases, although the main product might have its own autonomous profile, other collateral and intangible products can be identified in the assistance and consulting services provided to clients. The quality of such "products" and their consistency with client's expectations often significantly affect the main product performance.

The analysis of clients/products must be complemented by an internal one. This implies the identification of management processes, i.e., groups of homogeneous and inter-related activities, so as to generate a well-identified intermediate result oriented to the attainment of a final product.

Identifying a process system allows one to better focus the performance drivers related to the instrumental view.

In this regard, the concept of "product" is extended to what we can define as *intermediate product*, i.e., to the services provided by back-office units to their own *internal clients* (Fig. 3.43). Performance in delivering one or several sequential intermediate products will affect the performance of the *internal* clients receiving them, and will also influence the performance of other internal clients who are sequentially located along the value chain leading to the delivery of the *final* product (Bianchi 2010, p. 366).

---

[16]Though the "clients-products" framework may seem more compatible to an enterprise than to public-sector and non-profit organizations, the competitive viewpoint also can be effectively applied to such institutions. In fact, just as with businesses, so with a public institution: strategy is characterized by a set of goals implying the identification of stakeholders (e.g., community, citizens, users) for the satisfaction of whose needs a given 'service delivery system' is carried out. Similarly, just as with an enterprise, so too with a governmental institution: a set of strategic business areas can be defined, and the different units of service delivery (e.g., the policies carried out to satisfy groups of community or citizens' needs) which profile the strategic field where public institutions operate, can be outlined.

Fig. 3.43 Identifying a final
product and the related
intermediate products through
the objective view

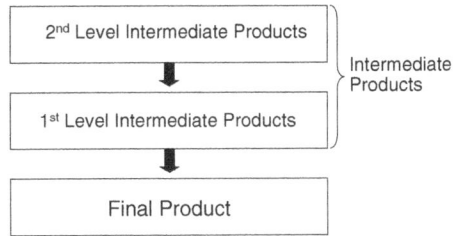

For instance, research and development acts as an internal service unit with
respect to marketing. So, the new product models that R&D first delivers to the
benefit of the second unit are intermediate products. Likewise, a purchase order or a
maintenance response is an intermediate product delivered by back-office units to
the benefit of their own internal clients.

A final product, which is delivered to an external client, can provide the basis for
the delivery of a set of other final products to the same external client (Fig. 3.44).

For instance, in a bank, a client's bank account number can be considered as a
"final" product, which allows the bank to deliver a package of related "final"
products, such as:

- Statements of account;
- Financial consulting;
- Financial transactions (e.g. bonds subscription);
- Sale of collateral services (e.g. ATM and credit cards, POS).

In a University, a Masters degree is a "final" product. However, in order to
deliver such product to students at the completion of their studies, a University must
also provide them with a wider final products "package", such as:

- Student ID number and enrollment certificate;
- Certificate of enrollment to the next year;
- Certificate of approval of the student's curriculum of studies;
- Transcript of records;
- The completion of internships or project works, or periods of study abroad (e.g.
  Erasmus programs).

Also in the public sector, an administrative product (Knoepfel et al. 2007, p. 206,
236) may take on a different connotation when seen as a function of the client[17] or
user to whom it is delivered. Of course, if one refers to the *external client*, i.e., to a
subject operating outside the public administration and receiving the outcome of
processes fulfilled by different public sector institutions in the *value chain*, a *final*
product can be identified.

---

[17]'Clients' are here meant as the subjects in the interest of whom each public sector unit delivers its
services.

Fig. 3.44 Identifying a final "product" package in relation to an external client

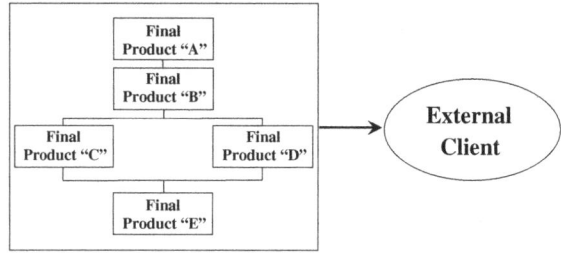

However, many public sector activities result in the delivery of intermediate products, to the benefit of units operating either inside the same institution or in other public institutions. Therefore, in relation to a final product, one can identify a system of intermediate products resulting from the fulfillment of processes by each decision unit whose clients are *internal* to the public sector.

Particularly in the public sector, understanding the impact of back-office units on delivered services is a necessary step toward improving performance and fostering accountability. This is not an easy task, because a bureaucratic perspective tends to be adopted when the contribution of such units to public service is considered (Osborne and Gaebler 1992, Chap. 6). Provided that back-office units take a major role in delivering such service (Millard et al. 2004), this perspective significantly tackles efforts to make public-sector decision makers accountable. Overcoming this problem requires the identification of 'administrative products', i.e. the measurable results generated by the tasks fulfilled by both back and front office units.[18]

This allows one to shift the view of public administration from a bureaucratic to a citizen and community-oriented one (Barzelay 1992, Chap. 8). In fact, performance targets can drive the activity of public-sector units and the evaluation of achieved results.

One implication of this analysis is that focusing objectives and performance measures exclusively on front-office units and related products delivered to external clients might well generate an unbalanced attribution of responsibility (Bianchi 2010. pp. 362–363). In fact, back-office units would not be held accountable for the impact of the results produced by their fulfilled activities on other units and so on the final products provided to external clients.

Therefore, the design of a performance management system requires that the chain of final and intermediate products delivered to both external and internal clients be fully mapped. It also requires that the underlying processes, responsibility areas, assigned resources, and policy levers be made explicit. These design requirements can be described as an objective view of performance management. As depicted in Fig. 3.45, such a view implies that products generated by management processes and activities are made explicit.

---

[18]The identification of administrative products is one of the building blocks of the so-called 'New Steering Model', i.e. a reform package introduced in Germany in the 1990s (Pollitt and Bouckaert 2000, p. 239).

Fig. 3.45 The "objective"
view of performance

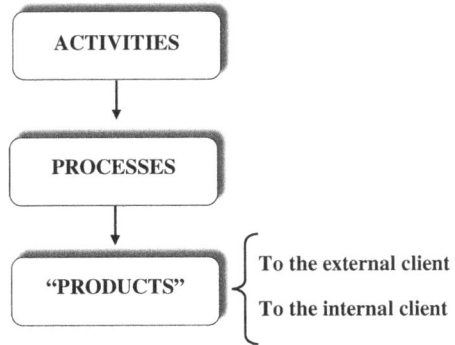

The identification of "products" and of the related macro-processes character-izing the objective view enhances the planners' ability to improve the focus of performance drivers that were initially outlined on a Corporate/Divisional and Departmental level, according to the instrumental view.

Figure 3.46 frames how the objective view is rooted on a departmental level. By adopting this view, one may understand not only what performance measures each department should affect in order to contribute to corporate performance, but also *how* this will be made possible. A proper level of synthesis through the objective view is provided by the identification of macro-processes.

Identifying macro-processes also supports a focused in cascading corporate performance drivers on an operational level, through the instrumental view (Fig. 3.47).

3.4.1   Mapping Products, Processes and Performance Measures in a Retail Bank: Case-Study

DPM has been used to map the product system, processes and performance indi-cators in a bank delivering a deposit service to so-called "Personal" clients (i.e.

Fig. 3.46 Focusing the
objective view on a
departmental level

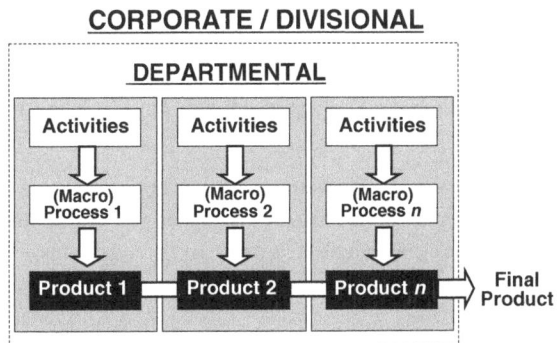

MACRO PROCESS 1   **MACRO PROCESS 2**   **MACRO PROCESS n**

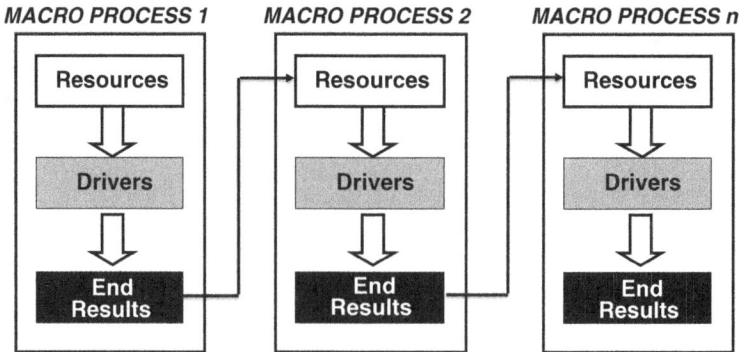

Fig. 3.47 From an objective to an instrumental view: cascading performance drivers on an operational level

whose savings are higher than €100,000). Such service is characterized by a system of inter-related sub-services, among which financial consulting is an important component. "Financial consulting" is a final product, delivered to the external client. In relation to such product, in a budget session, the Retail Division explores alternative strategies to attain a set of end-results, in terms of: (1) change in the customer base; (2) change in customer satisfaction, and (3) gross operating income from services to personal clients.

In respect to these end-results, the underlying processes through which financial consulting is delivered have been mapped. This allowed us to find out the drivers, strategic resources and policy levers to manage such resources, to affect financial and competitive performance.

Each branch of the retail division delivers such services. Within each branch, a Personal Banker/Consultant is made accountable on the achievement of the targets included in the operating budget at divisional level. To this end, the bank consultants outline a plan of meetings with the branch's Personal Clients, with the goal to analyze with each of them his/her current financial position and investment needs, based on four main kinds of needs, i.e.:

- Liquidity the client will have to hold in his deposits, to face short term financial expenditures;
- Speculative investments on potentially profitable financial tools, which do not provide any guarantees;
- Possible specific objectives (e.g., property purchase) to plan;
- The increase of invested capital, based on the expectation of profitability and risk.

After an analysis of financial decisions made by the client in the past has been done, the consultant estimates the customer's risk profile. Through an interview, the level of risk the client is willing to tolerate is assessed. For each client, the consultant tries to pursue an equilibrium point be-tween the objective to maximize yield and the need to minimize investment risks. In order to optimize the "Risk/Yield" ratio on a medium-long time horizon, the consultant uses an "efficient frontier" financial tool. Such tool supports the consultant in setting the pool of financial portfolios maximizing yield or minimizing risk, given a pre-defined yield. This is called the efficient frontier, i.e. the financial portfolio that the consultant proposes to the client, based on his risk profile and investment time horizon.

This implies that the consultant is always in contact with clients, and periodically visits them, to understand their needs and to examine the consistency of the allocation of their financial portfolio with market opportunities.

The macro-processes underlying the financial consulting service delivery to the Personal Client are: (1) interview; (2) analysis of the client's portfolio; (3) sale of financial products; (4) post-sale assistance. Each macro-process underlies the delivery of instrumental products, which are the results of the activities carried out by back-office units, in order to attain the final product ("consulting") to the benefit of the client. The different processes are outlined in Fig. 3.48.

Based on this analysis, the "products" generated by the listed macro-processes can be identified; planned/achieved results can be tracked, and responsibility areas can be made accountable on them. Figure 3.49 shows how the goal of the

Macro-process 1–Interview:
- Contact
- Setting the interview date
- Interview
- Customer profiling: risk orientation is analyzed.

Macro-process 2 – Exam of the client's investment Portfolio:
- Details on each single asset of the clients are inserted in a software to define the client's positioning on a "risk/yield" diagram, with respect to the "efficient frontier"
- A report positioning the client on the efficient frontier is printed
- The report is analyzed with the client
- A sale proposal is outlined.

Macro-process 3 – Sale of Financial Products:
- A follow up to the sale proposals is done through phone contacts
- A meeting with the client is agreed to propose a sale of financial products
- The client is met and products are sold
- Sale order data is inserted in the computer system
- The sale contract is printed and signed.

Macro-process 4 – Post-sale assistance:
- The market is monitored through periodic reports
- The client is periodically contacted
- The client's anxiety is managed
- New contracts are signed with post-sale assistance clients.

Fig. 3.48 Process detail in financial consulting delivery

MACRO PROCESS 1	MACRO PROCESS 2	MACRO PROCESS 3	MACRO PROCESS 4	
Contact with customers to profile	Customer positioning on the "efficient frontier"			2ND LEVEL INTERMEDIAT PRODUCTS
Customer profiling	Sale proposal	Setting up of a meeting with the client for sale proposal	Post sale assistance contacts	1ST LEVEL INTERMEDIAT PRODUCTS
Financial consulting contracts subscription				FINAL PRODUCT

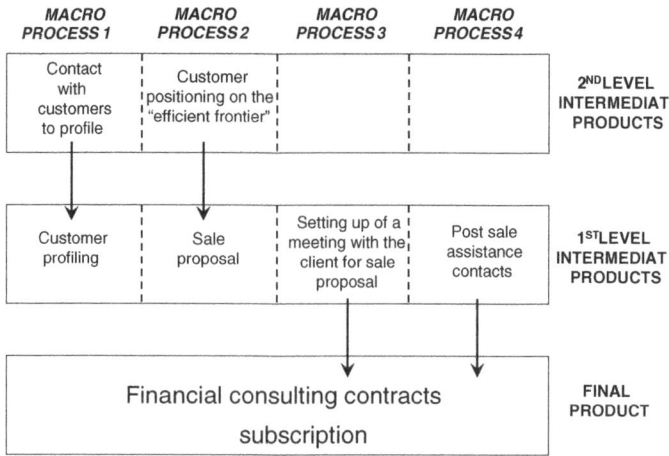

Fig. 3.49 Sequential product conceptual levels in final product delivery

macro-process-1 can be referred to as "customer profiling". This is an intermediate product, in respect to the final goal of the described processes, i.e., financial consulting contracts subscription.

The number of profiled clients is a performance measure that can be associated with such intermediate product. In order to affect this measure, a 2nd level intermediate product is identified, i.e., "contacts with customers to profile". A performance measure related to this product is the "number of interviews to profile clients".

Therefore, the identification of two levels of intermediate products supports corresponding performance measures. Regarding such measures, if we consider the Retail Division, the ratio between "Profiled clients" (actual performance) and "Clients to interview and profile" (planned performance), defines a "Profiling effort ratio". Since client profiling is a competitive success factor, this ratio is a driver of the Division's end-results. It affects the "Clients to contact loss rate", which in turn affects the "Change in the customer base". This "last layer" of divisional end-results provides a synthetic measure of competitive performance and determines the new level of the "Customer base" stock (Fig. 3.50).

Also, if we consider a single branch in the Retail Division, the described product and process analysis allows us to understand how to affect the profiling effort ratio. If we analyze process-1, we can observe how the number of profiled customers is an end-result for such process, and its driver is the "number of interviews to customers to profile". Such end-result is a flow in the "Profiled clients" strategic resource, which is used in process-2 (Fig. 3.51).

The same reasoning applies for the other macro-processes. So, regarding process-2, the sale proposal and customer positioning on the "efficient frontier" are first and second level intermediate products, respectively. Their corresponding performance measures are, respectively: the "sale proposal ratio" (i.e. between

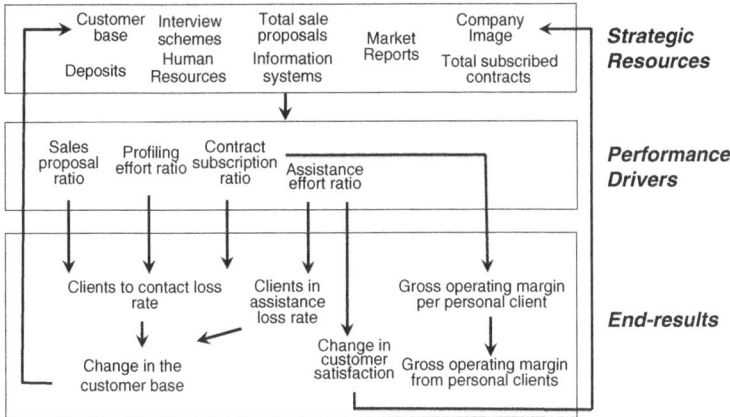

Fig. 3.50 Strategic resources, performance drivers and end-results on a divisional level

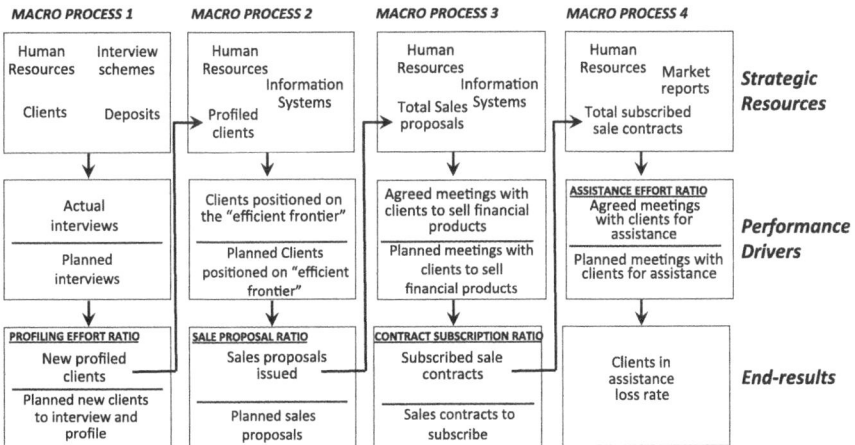

Fig. 3.51 Strategic resources, performance drivers and end-results on a process/branch level

actual and planned sales proposals) and the ratio between the actual and planned number of clients positioned on the "efficient frontier". If observed on a divisional perspective, the "sales proposal ratio" is a driver of the "clients to contact loss rate".

Since process-3 and process-4 are closer to the final "product" delivery, only one level of intermediate products has been outlined. Such products are, respectively related to the "meetings setting up for sale proposal" and to "post sale assistance contacts".

More specifically, concerning process-3, corresponding performance measures are, respectively: the "contract subscription ratio" and the ratio between the actual and planned number of meetings with clients to sell financial products (volume measure). If observed on a divisional perspective, the "contract subscription ratio" is again a driver of the "clients to contact loss rate". It also affects the gross operating margin per client, which in turn affects the total gross operating margin flow from "personal" clients.

Concerning process-4, a performance measure corresponding to the detected intermediate product is the "assistance effort ratio" (i.e. between actual and planned meetings with clients). If observed on a divisional perspective, the "assistance effort ratio" is a driver of the "clients in assistance loss rate", which affects the change in the customer base. The "assistance effort ratio" also affects the change in customer satisfaction (end-result), which affects the company image stock.

To show the usefulness of SD in mapping the described products, processes, and performance measures, a simple stock-and-flow model will be discussed now. The model structure will not be described with the purpose to identify alter-native policies from problem behavior. The purpose is, rather, to show how SD can add rigor to performance analysis by implementing the conceptual framework here discussed. The value SD adds to the design of performance management systems refers to a proper identification of: performance measures, strategic resources, policy levers and performance standards, and budget objectives.

The model portrayed in Fig. 3.52 describes the four macro-processes previously commented. Through a stock-and-flow chain it identifies the factors affecting financial consulting performance management. The unit of measure of stocks is "clients"; the unit of measure of flows is "clients/time".

Performance standards are a percentage of maximum potential workload. For instance, in "process-1", to get the budget objective "Clients to interview and profile", the "% STD Clients to profile" is multiplied by "Total Clients Profiling" (i.e. the sum of the three stocks in the process).

Performance drivers are ratios between a stock representing a delayed infor-mation value of the "product" volume resulting from each process and the corre-sponding objective. So, drivers (e.g. the "Profiling effort ratio") are expressed in relative terms. The effect that each driver generates on corresponding end-results is gauged through normalized graph functions, which take the driver as an input. Such functions are expressed as multipliers. For instance, the driver named "Effect of profiling on customer loss" is a value comprised between 0 and 0.2. It represents a percentage of the "Customers to contact" stock which is drained as a client loss rate in a given time.

The "Clients to contact loss rate" is also affected by two other multipliers, i.e.: "Effect of subscribed contracts on Customer loss" and "Effect of sale proposal on Customer loss". The performance drivers affecting the two multipliers are respec-tively originated in process-2 ("Sales proposals ratio") and process-3 ("Contract subscription ratio").

The other performance driver, originated in process-4 ("Assistance effort ratio"), affects the "Clients in assistance loss rate". It is a ratio between the number of agreed meetings with clients in assistance, and the corresponding number of clients to assist in a given time.

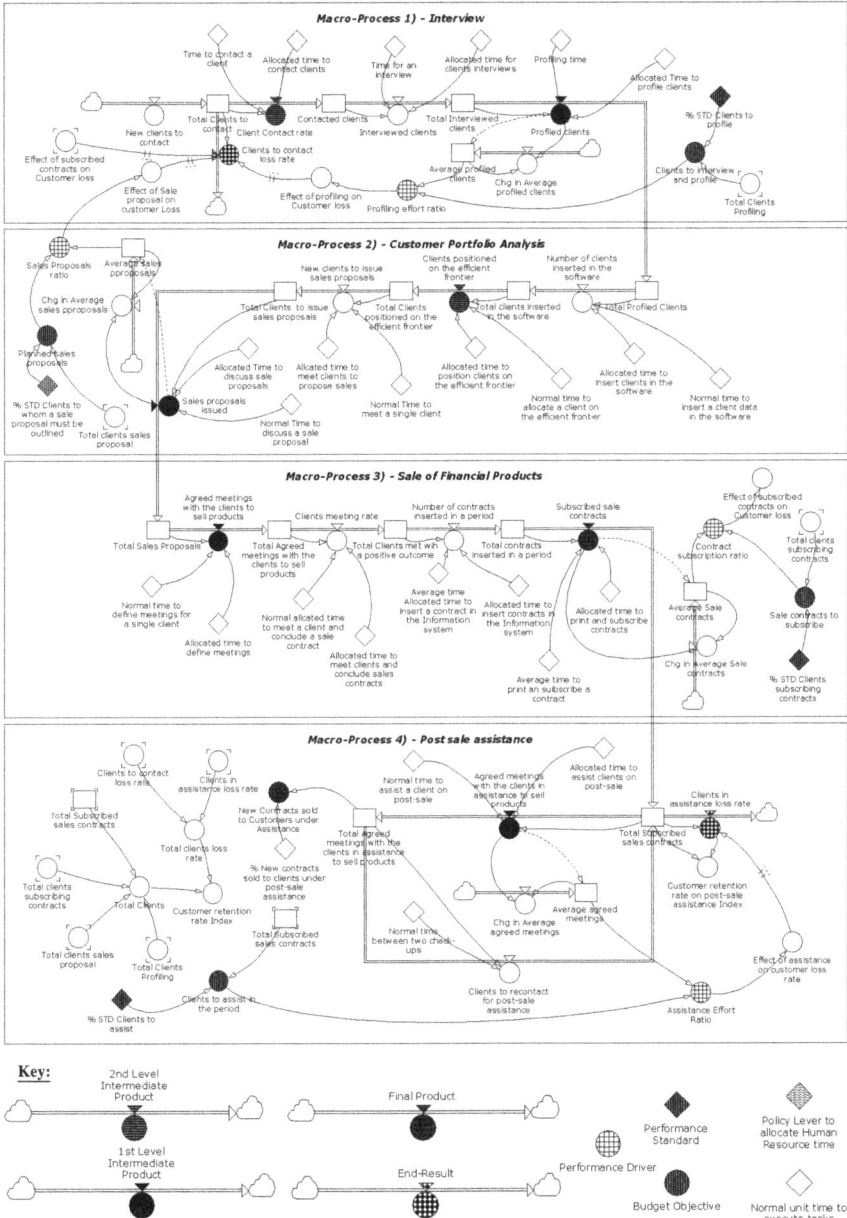

Fig. 3.52 A stock-and-flow depiction of products, processes, and performance measures

SD modeling also contributes to dynamic performance management since it helps managers to avoid common errors in BSC practice in the identification of causalities between measures. For instance, in static BSCs, performance indexes are often confused with performance drivers. An example is provided here by the customer retention rate. This variable is a synthetic expression of performance, which cannot be confused as a driver in this context. In fact, it does not affect the customer loss rate. It is rather an effect of it.

3.4.2 Mapping Products, Processes and Performance Measures in a Public Utility: Case-Study

A second case will be now illustrated to show how the objective view has been applied in a public utility water company.

In this company, not only the water provided but also the billing service can be considered as a 'product' according to which external clients may have a number of expectations. They may expect that billing be done according to proper accuracy, transparency and reliability standards. Therefore, billing errors and irregular updates of consumptions will result in a poor service, though the quality, availability or cost of supplied water is satisfactory. Likewise, the billing 'product' will be considered as poor, if uncollected invoices—due to unreliable billing—will reduce the public utility cash flows.

Related to billing (as a final 'product'), relevant 'intermediate products' for such organization, could be referred to: meters inspection, billing issuing, checking billing errors and compulsory credit collection.

Although most of such 'products' are delivered by back-office units, they have a direct impact on the delivered final product (i.e. billing), resulting in a higher or lower customer satisfaction and utility solvency.

Most of the above 'intermediate products' are delivered to 'internal clients'. For instance, the performance of a 'customer complaints office', receiving requests from clients, due to bad invoicing and/or meters checking, are affected by the speediness and accuracy of activities carried out by various back-office units, such as those fulfilling operations related to the inspection of the correct recording and billing of consumed water.

This analysis suggests the need to track management processes and underlying activities, in order to explain how to improve results associated with the 'products' delivered to both internal and external 'clients' related to a given public sector organization.

All activities pertaining to the periodical check of meters/bills issuing, aimed to assess consumptions, can be grouped into a process, whose result (i.e. meters inspection) affects the accuracy and predictability of billing. The water utility company's 'services charter' prescribed that the average time between two inspections should have been 6 months. Since in this case, due to an excess of workload on the staff in charge of checking meters, this time was increasing, in spite of the automatic resort to overtime, this caused a poor service. In fact, for those customers whose meters were not regularly checked, billing was done based on past water consumption.

Therefore, actual consumption was randomly updated and many clients found significant unexpected and unpleasant extra-charges in bills, which were instead punctually issued every 2 months. This phenomenon increased the percentage of uncollected bills and their collection time.

Likewise, also those activities regarding the assessment of billing errors influenced the bills collection time. Such errors were increasing because of a rising resort to overtime for checking meters: in fact, increasing overtime implied a higher occurrence of errors by the company staff in checking actual consumptions.

In order to reduce the disputed bills collection time, a process similar to the one previously noted (checking meters) was carried out. Precisely, a group of senior inspectors was requested to verify possible errors in checking meters, and—when necessary—to start the procedure for re-issuing the wrong bills.

In the utility, different areas had a joint responsibility for the fulfillment of the previously described processes, i.e.: the periodical check of meters/bills issuing, and the assessment of billing errors.

Activities underlying the above two processes were managed by three units, i.e.: Commercial, Legal and Technical. The Commercial unit also consisted of five offices: Customer complaints, Contracting, Invoicing, Meters installation and Electronic Data Processing (EDP).

The invoicing process was started by periodical meters inspections by the Invoicing office. The same office was also responsible of sending senior inspectors to check meters in case of customer complaints about suspected billing errors. It was also responsible for asking the 'Meters installation' office to replace damaged meters with new ones. Once water consumption was periodically ascertained, bills were issued automatically by the EDP.

For those bills which had not been collected after 2 months from their issue, an overdue debt was reported after 2 months in a new bill, together with the new accumulated debt in the last period. If, after two more months, the accounts receivable had not been collected yet, the 'Invoicing office' would have communicated the total overdue to the 'Credit collection office' of the Legal unit. This last office was used to send a letter to the client, to intimate payment of the overdue, including accumulated interest. After receiving such letter, most clients were used to complain about wrong billing at the 'Contracting' and 'Customers' offices. Some of them were also used to ask the company to test the correct functioning of their meter. This implied that the 'Invoicing office' had to check meters again, in order to verify possible billing errors; in some cases the 'Technical office' was also asked by the 'Meters installation' office to replace old meters with new ones (Fig. 3.53).

These messy problems led the company to three main consequences: (1) a liquidity crisis, due to the long delays in collecting overdue bills; (2) strong conflicts between different units and offices, all of which felt themselves not liable for the recorded inconveniences; (3) a low customer satisfaction, due to a lack of confidence towards the utility about the precision and reliability of recorded consumed water.

PERIODIC METERS CHECKING	BILLING	COMPLAINT RECORDING	CHECKING BILLING ERRORS
Customer Billing Office • Checking meters • Recording consumption • Input data to CED	*CED Office* • Processing data • Invoice issue • Invoice dispatching and notification of overdue • Sending the overdue list to the Legal office	*Public Relations Office* • Checking status through terminal • Supporting customer in filling the 'complaint form' requiring a check of the meter • Forwarding the complaint to the 'Billing' office	*Billing & Public Relations Offices* • Checking meters • Consumptions are communicated to the PRs' Office • The PRs' Office communicates the customer results from checking

COMPLAINT

INVOICE COLLECTION

MANAGING AND COLLECTING OVERDUE	CHECKING METERS' FUNCTIONALITY
Legal Office • Checking overdue • A letter of overdue is sent to customers • Compulsory credit collection procedures are started • A supply suspension request is sent to the Connections Office	*Connections Office & Technical Service* • A technical team is asked to remove the meter • A new meter is installed • The old meter is sent to the Technical Service to check its functionality • The new meter's code is communicated to the Billing office

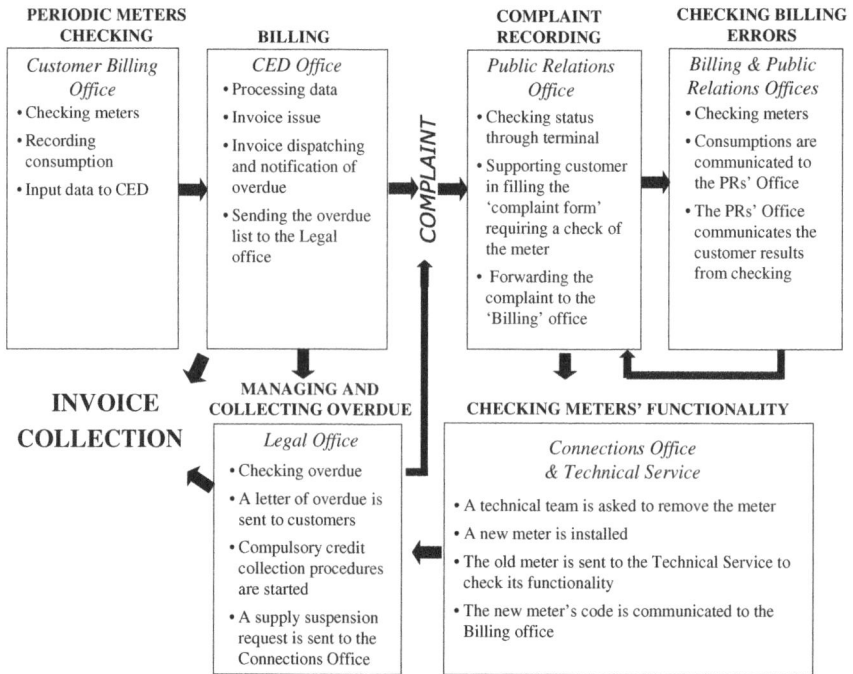

Fig. 3.53 Main processes and organization units responsible for the 'billing' product in a water public utility

Although possible 'solutions' might seem obvious, if we analyze the problem from outside the public utility and after events have taken place, the organization was not able to perceive reality.[19]

Figure 3.54 shows that the company policy was to respond—after a delay—to a rise into bills collection time by increasing staff overtime devoted to both checking meters and uncollected billing errors. In the short term, for both processes, increasing overtime allowed the utility to improve (i.e. to reduce) two important performance drivers: meters inspection time and time to check billing errors. This, in turn, allowed the utility to increase its output (in terms of checked meters and bills, respectively) and to decrease the bills collection time (main performance driver affecting cash flows).

However, over a longer time span, the 'uncollected bills' problem emerged again. In fact, overtime productivity was gradually decreasing, due to burnout. This was a

---

[19]This phenomenon is due to a lack of coordination and communication between the different (back and front-office) units, and to a poor perception of delays. SD modeling can play an important role in dealing with these problems, and fostering a process—rather than function—oriented view of performance. A process-oriented view implies that each organization unit is made accountable on a set of indicators pertaining to the 'products' resulting from each process to which it contributes. It also implies that, for each process, the impact of other units on results, and the effects generated by material and information delays are taken into account.

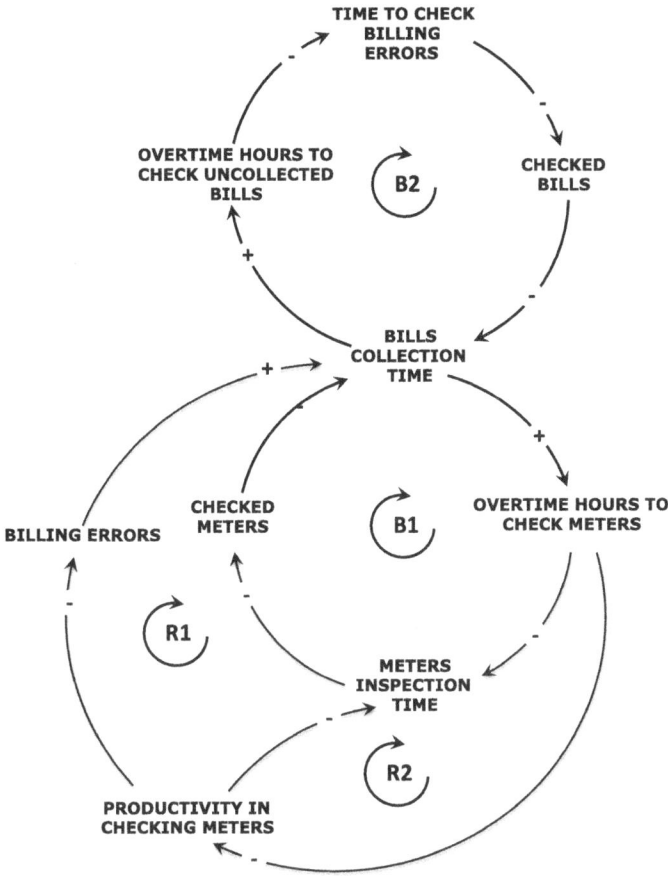

Fig. 3.54 Unintended effects of overtime policies in a public water utility. Loop "B1" and "B2": symptomatic solutions to service problems ('meters inspection time' and 'time to check billing errors', respectively); Loop "R1" and "R2": unintended effects amplifying service problems over a longer time span (billing errors and meters inspection time, respectively)

major cause of more billing errors and a lower number of meters checked per hour, which contributed to rise the meters inspection time again. Therefore, both phenomena caused a new increase in the billing collection time. So, managers were forced to periodically resort to overtime, but this did not allow them to solve the problem.

Figure 3.55 depicts main performance drivers' dynamics produced by the above policy, as simulated through a SD model developed with the collaboration of a manager of the utility, operating in the budgeting unit.[20]

---

[20]The model was a follow-up of a previous DPM project based on a BSC embodying detailed company data (Bianchi and Montemaggiore 2008). The aim of modeling was to analyze more in depth a number of processes and inter-relationships that—in order to follow a same level of

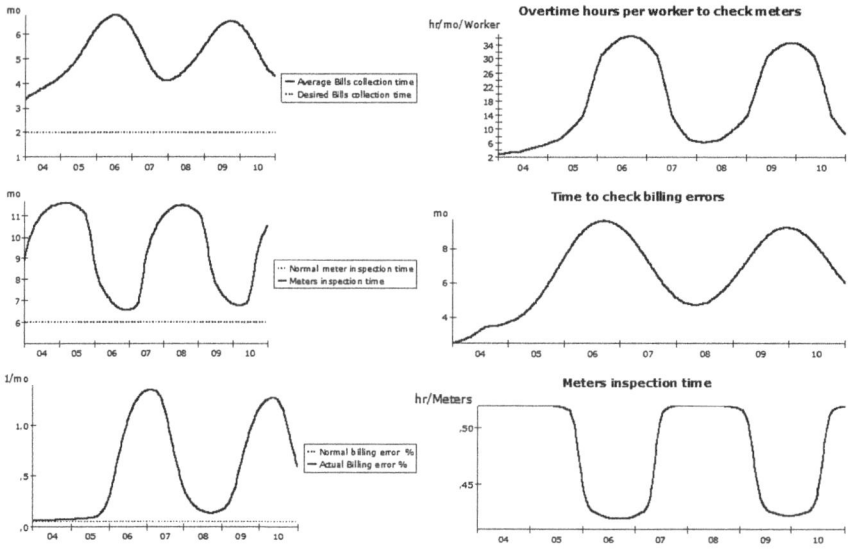

Fig. 3.55 Main performance drivers' dynamics from overtime policies in a public water utility

The periodic oscillations portrayed by the graphs demonstrate the structural inability of the above policy to ensure company performance stability in the long run.

Simulation helped the utility managers to detect major policy weaknesses and to envisage possible solutions to the experienced problems.[21]

Figure 3.56 shows main feedback loops related to a sustainable policy. The balancing loops B1 and B2 identify the number of needed workers (strategic resource) to allocate to meters inspection and checking billing errors, in order to meet performance targets. Based on the desired number of workers, the suggested policy aims to fill human resource gaps (balancing loops B3 and B4).

The feedback structure depicted in Fig. 3.56 was embodied in a new version of the simulation model, in place of the feedback structure illustrated in Fig. 3.54. Simulation results demonstrated the sustainability of this new policy.

(Footnote 20 continued)

analysis for the different subsystems—were not included in the BSC model. However, this was an insight model, including—as company real data—the number of bills, the real behavior of billing collection time and the associated performance drivers, as well as main delays characterizing the investigated company processes. The model was used to stimulate a dialogue between the involved units and a better understanding of their role in affecting performance.

[21]For instance, it was remarked that a possible fundamental solution could have been to increase the number of inspectors devoted to check meters. If necessary, also senior inspectors could have been devoted to periodical checking, in order to reduce the backlog of meters to check, so to keep the company on track with the '6 months' average inspection time target. This would have allowed the utility to reach a more reliable level of billing accuracy.

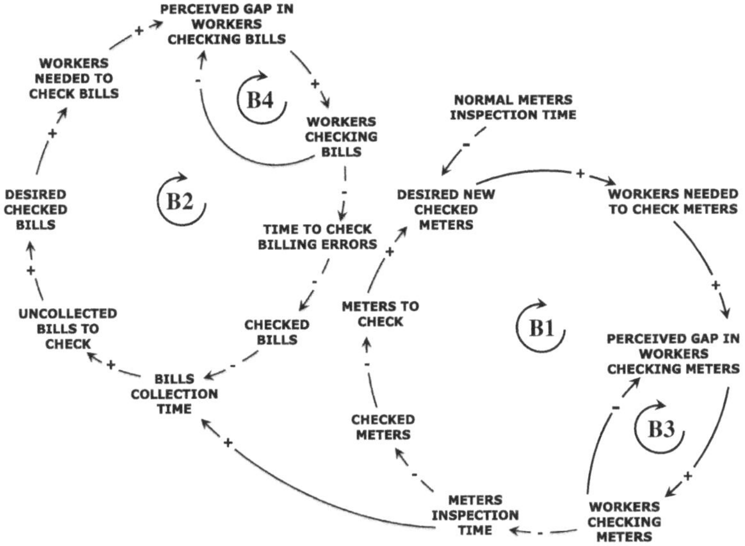

Fig. 3.56 An alternative sustainable policy to fix the billing errors and collection time delays in a public water utility. Loop "B1" and "B2": fundamental solutions to service problems ('meters inspection time' and 'time to check billing errors', respectively); Loop "B3" and "B4": strategic resource adjustments sustaining fundamental solution to service problems (hiring needed workers gap in checking meters and checking billing errors, respectively)

Figure 3.57 shows how performance drivers were now showing a more stable behaviour, which was proving the sustainability of the policy.

Decision makers in this public utility were not able to perceive the causes of experienced oscillations in performance targets. This was because of the dynamic complexity of the underlying system: more then 20,000 invoices were issued per month and about 40,000 m had to be checked on average twice a year. Delays between a step of the billing/collection process and the next one, as well as interdependencies between different responsibility areas, and a missing process-oriented vision, were also significant factors inhibiting decision makers to properly frame such an apparently simple and predictable system. The above problems were also increased by the difficulty of the company management/ strategic control to perceive small performance changes.

Figure 3.58 provides a synopsis of how the main SD model's stock-and-flow structure supported the utility in detecting the causal relationships between 'products', processes, and strategic resources on which to act in order to affect performance drivers and end-results, to foster accountability. Strategic assets are depicted as stocks primarily affecting the system's performance. End-results are modeled as flows changing strategic resource endowments. Performance drivers are auxiliary variables. Processes come from the analysis of factors impacting on the flows

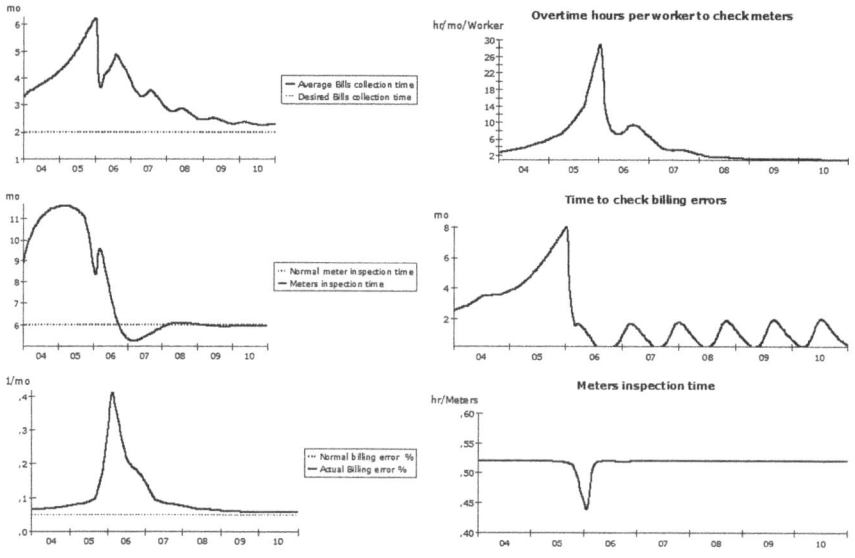

Fig. 3.57 Main performance drivers' dynamics from the adoption of a sustainable policy to fix the billing errors and collection time delays in a public water utility

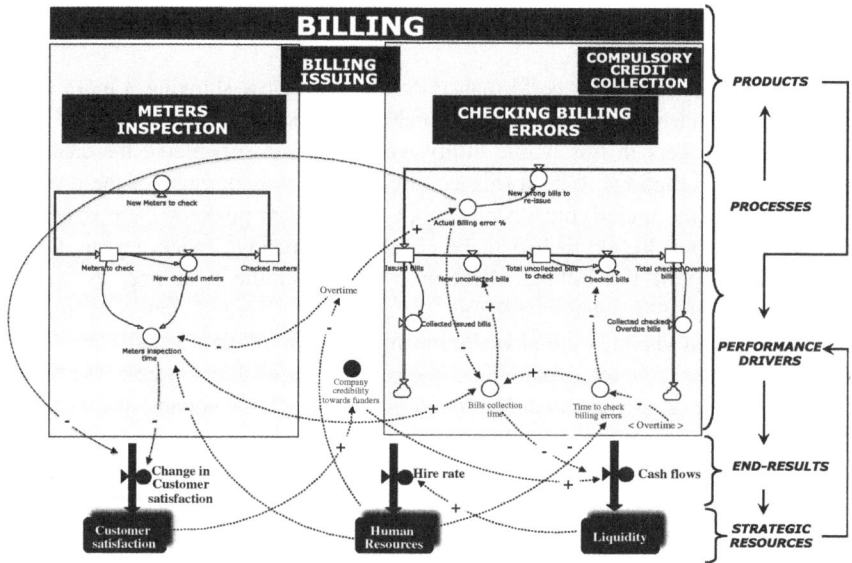

Fig. 3.58 A synthetic picture of the main feedback stock-and-flow structure supporting DPM in a public water utility

affecting stocks to which performance drivers are related (i.e. those concerning meters and bills). Products are an outcome of the carried-out processes.[22]

3.5   The Subjective View of Performance

The "subjective" view provides a synthesis of the instrumental and the objective view, because it makes explicit, as a function of the pursued results, both the activities to undertake and the related objectives and performance targets to include in plans and budgets for each decision area. This view requires that performance measures—i.e., drivers and end-results—associated with the delivery of products are made explicit, and are then linked to the goals and objectives of decision-makers (Fig. 3.59).

Results originate in decision-makers' activities as those are carried out in the various fulfilled processes. Expected results provide a benchmark to which to refer for setting goals and objectives for each responsibility area in a P&C context. Both objectives and performance measures can be gauged on a corporate and strategic business area level, or in the public sector on the level of governmental functions.

Both are related to the expected end-results and performance-driver targets assigned to the various responsible areas in a firm, within a budgeting context.

Therefore, activities and the processes to which they are related can be associated with corresponding objectives and performance measures, in a consistent action plan, from which resources are assigned in an organization, available policy levers for each decision area are made explicit, and responsibility for expected results is focused. A remaining crucial issue concerns the need to properly identify performance measures so as to assign to decision areas for performance evaluation in a budgeting and control process.

In this regard, the need for specific, measurable, achievable, relevant, and time-related objectives (Conzemius and O' Neill 2006) has been emphasized. Also

---

[22]Figure 3.58 is a synthetic picture of the SD model that was developed with the utility manager's support. The dotted links it embodies represent logical relationships between variables; therefore, some of them are not included in the model in the same way they are presented here. For instance, concerning the 'meters inspection process', in the simulation model the variable 'Productivity' has been represented through an auxiliary variable named 'Number of meters checked in an hour'. This variable (multiplied by the number of worked hours) determines the 'Checked meters' flow. From the ratio between the 'Meters to check' stock and the above said flow, the model calculates the auxiliary variable named 'Meters inspection time'. In the simulation model, this variable, in turn, affects the bills collection time through the stock-and-flow chain regarding uncollected bills (which is longer than the one depicted in Fig. 3.58). In fact, it determines the 'Collected issued bills' flow, which affects the 'New uncollected bills' flow, which accumulates into the 'Total uncollected bills' stock. The model calculates the bills collection time as a weighted average between four major delays affecting bills collection, i.e.: (1) issued bills collection time (ratio between "issued bills" and "collected issued bills"); (2) average time to check uncollected bills; (3) overdue bills collection time and (4) overdue bills forwarded to the legal department collection time. The stocks of bills at each stage of the described processes have been used as weighting factors.

Fig. 3.59 The "subjective"
view of performance

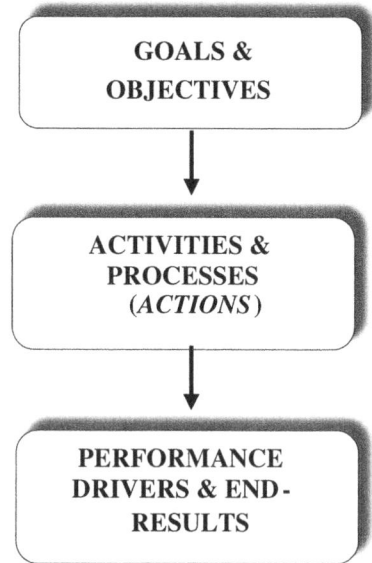

the risks associated with improper goal-setting have been analyzed, in particular by focusing on behavioral distortions that it could generate for decision-makers. Such phenomena are connected to:

- Unfocused goals, leading managers to maximize their own efforts towards a subset of the overall relevant picture (Merchant 1997, pp. 454–541);
- Bounded attention towards non-monetary goals, leading managers to focus their own decisions only on improving financial results, rather than also on qualitative factors impacting on performance;
- A distortion between means and ends, leading to an exclusive focus on the constitution of resources, rather than also on their effect upon performance;
- A deliberate downgrading of performance standards, against which actual performance levels will be compared when the performance cycle will be closed.

To provide the reader with a more practical understanding of how the methodological framework outlined here has been applied in practice, the next chapters will outline further examples and case-studies through which DPM will be applied.

3.6   An Integrative Framework of Performance

The three performance views described here play a complementary role in a DPM system. In fact, the objective view defines what the object of performance management is. The instrumental view identifies how to affect the defined object(s). The

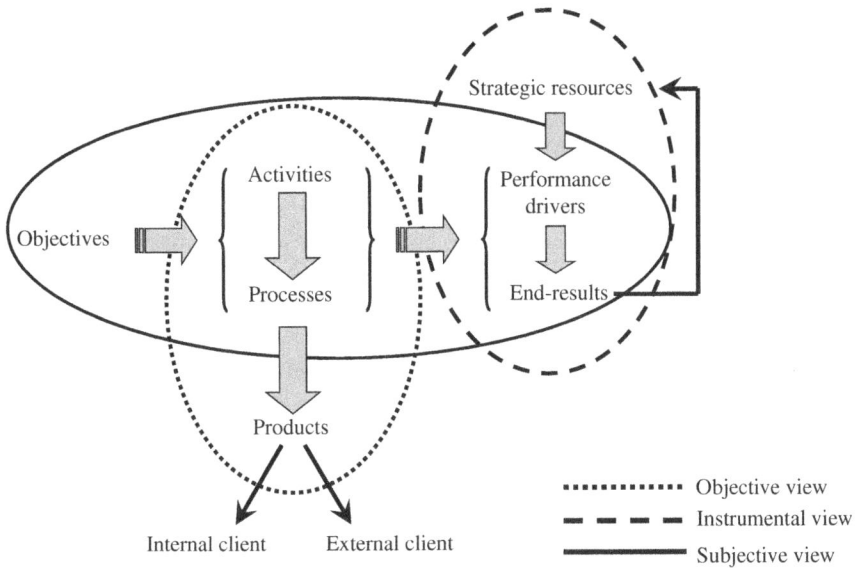

Fig. 3.60 Synergies among the three views of dynamic performance management

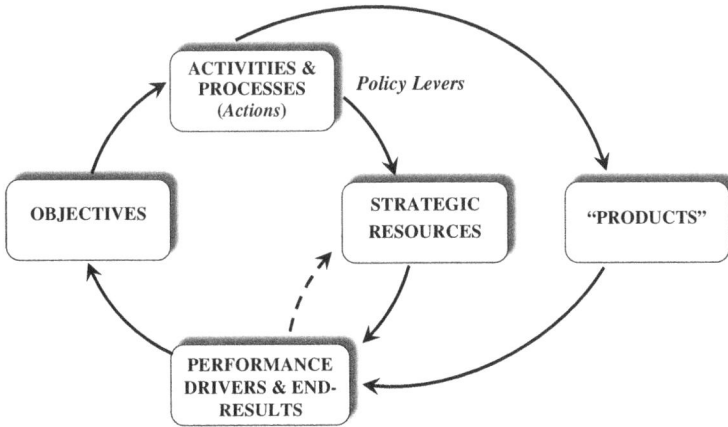

Fig. 3.61 The instrumental, objective and subjective views of DPM in action: a learning-oriented approach

subjective view focuses on who is responsible for the accomplishment of activities aimed at building and coordinating strategic resources, to affect performance drivers and end-results, and to obtain an estimate of the volume and quality of products/services so as to efficiently and effectively satisfy the needs of target clients.

Therefore, the interplay between the three performance views may support a responsive and learning-oriented P&C (Fig. 3.60). This is particularly crucial when organizations operate in dynamic complex systems, where a formalistic, static, non-systemic and incremental approach to planning has proved to generate an illusion of control leading to unsustainable growth and crisis (Brews and Hunt 1999; De Geus 1988; Mintzberg 1993, 1994a, b).

Figure 3.61 sketches a synthetic picture of the three previously described perspectives. It shows how, in a planning context, once the administrative products have been defined, it is necessary to move backwards, i.e., to outline the underlying processes and activities, and then to define goals and objectives for each responsibility area. Such objectives must correspond to the results and indicators that will be achieved through actions aimed at managing a given strategic-resource system.

Both performance drivers and end-results should describe whether an organization is able to meet the various expectations (e.g. in terms of volumes, defects, time, and cost) coming from internal and external clients, concerning delivered products.

References

Ammons, D. (2001). *Municipal benchmarks*. Thousand Oaks, CA: Sage Publications.

Barzelay, M. (1992). *Breaking through bureaucracy*. Oxford: University of California Press.

Bianchi, C. (2010). Improving performance and fostering accountability in the public sector through system dynamic modeling: From and 'External' to an 'Internal' perspective. *Systems Research and Behavioral Science, 27*, 361–384.

Bianchi, C., & Bivona, E. (2002). Systems gear (A) and (B). *CED4-system dynamics group—University of Palermo* (unpublished case-study available upon request from: bianchi.carmine@gmail.com).

Bianchi, C., Cosenz, F., & Marinkovic, M. (2015). Designing dynamic performance management systems to foster SME competitiveness according to a sustainable development perspective: Empirical evidences from a case-study. *International Journal of Business Performance Management, 16*, 1.

Bianchi, C., & Montemaggiore, G. (2008). Enhancing strategy design and planning in public utilities through "Dynamic" balanced scorecards: Insights from a project in a City Water Company. *System Dynamics Review, 24*(2), 175–213.

Brews, P. J., & Hunt, M. R. (1999). Learning to plan and planning to learn: Resolving the planning school/learning school debate. *Strategic Management Journal, 20*, 889–913.

Conzemius, A., & O' Neill, J. (2006). *The power of SMART goals: Using goals to improve student learning*. Bloomington: Solutions Tree.

Cooper, R., & Kaplan, R. (1988). How cost accounting distorts product costs. *Management Accounting, 69*(10), 20–27.

Cosenz, F., & Bianchi, C. (2013). Designing performance management systems in academic institutions: A dynamic performance management view. In *ASPA Conference*, New Orleans.

CRIBIS D&B. (2014). Payment Study 2014. Available from: https://www.cribis.com/media/78511/payment-study-2014.pdf

De Geus, A. (1988). Planning as learning. *Harvard Business Review, 66*(2), 70–74.

Hall, R. (1976). A system pathology of an organization: The rise and fall of the old Saturday Evening Post. *Administrative Science Quarterly, 21*(2), 185–211.

Knoepfel, P., Larrue, C., Varone, F., & Hill, M. (2007). *Public policy analysis*. Bristol: The Policy Press.

Maciariello, J. A. (1984). *Management control systems*. Englewood Cliffs: Prentice Hall.

Merchant, K. (1997). *Modern management control systems*. Upper Saddle River: Prentice Hall.

Millard, J., Svava Iversen, J., Kubicek, H., Westholm, H., & Cimander, R. (2004). *Reorganisation of Government Back Offices for Better Electronic Public Services—European good practices*. Main Report, Danish Technological Institute & Institut fur Informations-management GmbH, University of Bremen. http://ec.europa.eu/information_society/activities/egovernment/library/efficiency/index_en.htm

Mintzberg, H. (1993). The pitfalls of strategic planning. *California Management Review, 36*(1), 32–45.

Mintzberg, H. (1994a). Rethinking strategic planning part I: Pitfalls and Fallacies. *Long Range Planning, 27*(3), 12–21.

Mintzberg, H. (1994b). Rethinking strategic planning part II: New roles for planners. *Long Range Planning, 27*(3), 22–30.

Morecroft, J. (2007). *Strategic modeling and business dynamics*. Chichester: Wiley.

Osborne, D., & Gaebler, T. (1992). *Reinventing Government*. Reading: Addison-Wesley.

Pollitt, C., & Bouckaert, G. (2000). *Public management reform: A comparative analysis*. Oxford: University Press.

Richardson, G. P., & Pugh, A. L. (1981). *Introduction to system dynamics modeling*. Portland, Oregon: Productivity Press.

Schwaninger, M. (2009). *Intelligent organizations*. Berlin: Springer.

Warren, K. (2008). *Strategic management dynamics*. New York: Wiley.

Winter, S. (1964). Economic natural selection and the theory of the firm. *Yale Economic Essays, 4*, 225–279.

Wynne, J. (1985). The Saturday Evening Post—Revised. *Harvard Business School Case*, n. 9-373-009 (pp. 1–23). Boston: HBS Case Services, Harvard Business School.

Chapter 4
Applying Dynamic Performance Management to Public Sector Organizations

4.1 Introduction

This chapter will illustrate different examples of DPM applied to public sector organizations.

As remarked in Chap. 2, applying SD to foster performance management and to support effective decision-making in the public sector implies a number of challenges, which are related to the specific complexity of this environment (Bianchi 2010).

Such complexity can be referred to four main interdependent sub-systems (Borgonovi 1996, p. 105), i.e.:

1. Socio-economic system;
2. Political system;
3. Institutional system;
4. Organization system.

The design and implementation of programs aimed at fostering change, accountability and the improvement of service to users, and to the wider community, could be addressed by focusing predominantly on the organization system (e.g. on processes, roles and responsibilities, planning and reporting, career and reward/incentive schemes), and at the same time be hampered by a lack of attention to the wider socio-economic, cultural/political, and institutional systems with which an organization interacts in the public sector. Although such factors may be considered outside the scope of a traditional organizational intervention, ignoring them may cause a failure in reaching the objectives stated. Such a failure may arise from relying on the hypothesis that improving the organization system itself will cause an improvement in the overall performance as well. However, this hypothesis may not be valid for a number of reasons.

First, a public sector organization is connected with other independent (loosely coupled) public sector organizations, whose policies may affect (positively or adversely) the implementation of its organizational change programs (Bianchi

© Springer International Publishing Switzerland 2016   141
C. Bianchi, *Dynamic Performance Management*, System Dynamics
for Performance Management 1, DOI 10.1007/978-3-319-31845-5_4

2010). The rules and legislative schemes that define such relationships can be referred to as the institutional system that deals also with relationships between the public administration system and external actors (e.g. citizens and private sector organizations).

Second, rules of conduct are not only 'coded' by the formal institutional system; in fact, an informal political system defines how power is actually exerted and how the different roles that are ruled by the institutional system are played. This is affected by the hidden (social) rules of conduct leading to set political priorities in the allocation of resources, and to outlining community goals. Ignoring such a political system may significantly tackle any change program that only takes into account the managerial view of a public sector organization.

Third, since both the institutional and political systems are an expression of the emerging (or leading) culture and values arising from the underlying wider socio-economic systems, any public sector organizational change program ought to take into account what opportunities there exist to affect, in the long term, such a culture and values, possibly with the aim of changing the context (i.e. both the institutional and political system) in which the management operates.

A number of challenges can be envisaged regarding the possibility to support public sector performance management systems through SD modeling, i.e.:

– How can DPM facilitate a better understanding of the (often hidden) relationships between the political and managerial system?
– How can DPM foster coordination between the two levels? How can it foster a better communication and strategic dialogue between politicians and managers, with the intent to improve policy-making and implementation?
– How can DPM support management to keep under control the key performance variables driving to the success or failure of their organizations?
– How can DPM support policy makers to implement outcome-based performance management, and to improve social reporting on the long-term effects generated by implemented strategies?
– How can DPM improve coordination between different public and private sector institutions operating in a local area?

The following sections will illustrate different cases and topics in relation to which these problems have been framed.

4.2 Applying Dynamic Performance Management to Public Utilities: Water Provisioning, Distribution, and Wastewater Treatment

AMAP runs the municipal water provisioning and distribution service since 1950. About one million people live in the area it serves.

With the intent of fostering public utility efficiency and effectiveness, in the last decade the national government implemented a set of reforms. In particular, in 1994 the management of water resources was reorganized in order to avoid waste and to improve service quality.

Government regulations have been merging the sewer and wastewater treatment management with city water provisioning and distribution management, making all the municipal water service companies handle the so-called "integrated water cycle". In addition to this business re-engineering process, the regulator introduced competition for the management of the water service. This led to a privatization process, which implied the transformation of city water companies from public agencies to joint stock companies.

In this new scenario, the regulator assigns the water management service for a specific area to the company with the highest effectiveness (in terms of service quality) and with the best efficiency (in terms of service costs).

The changes in the water provision service rules made AMAP perceive the need to improve its performance in terms of both financial end-results and quality of the service supplied to customers.

Public utilities like AMAP, being joint stock companies, have a strong formal independent status. On the institutional viewpoint, an independent Board of directors outlines the company vision and main strategic goals. However, being the Municipality the main shareholder of the utility, political parties take a leading role in designating the members of such Board. This turns the utility's external autonomy into a formal state only.

Members of the Board are often appointed on the basis of their 'political affiliation' and rarely on their professional experience.[1]

In order to recover negative financial performance, Italian public utilities cannot autonomously set the water tariff. This is determined by a national law, based on the investment plan adopted by the public utility, so to allow the utility to earn a 7 % annual return on capital invested (Bianchi et al. 2010).

Among the internal factors affecting public utilities' performance, the degrees of internal autonomy and accountability, and customer orientation can be taken into account.

The strategic decision making process in AMAP was highly centralized. Although the company is organized in different departments, each head of these units had a limited decision power and very bounded autonomy. The Board mainly focused on strategic decision-making and the managing director focused only on operational decisions.

The decisions that were typically made by the managing director, and authorized by the Board, were those related to the approval of the procurement of goods and services and the hiring of employees. As a consequence, those politicians who appointed the board members could influence such decisions.

---

[1]This phenomenon is consistent with research conducted by Gutiérrez and Menozzi (2008) on 114 Italian public utilities in a 10-year period. In particular, the study demonstrates that unnecessary staff is often hired, because of political pressures on the Board, and such decisions have a negative impact on utilities' performance.

So, the company was lacking a shared vision of the mission as well as a coherent strategy for its accomplishment. Furthermore, communication between the different levels of the organization was almost absent and just a few of the middle managers and line workers were aware of the company's overall performance.

The low level of company internal accountability and transparency can also explain the high level of influence of political parties through the Board on the utility's strategic decisions. For instance, the company did not adopt a robust system to assess and manage organizational and individual performance.

Although some managers were used to undertake employees' evaluations at their own initiative, these evaluations were not structured. The formal respect of procedures was perceived as more important than meeting performance targets.

Furthermore, AMAP had a weak orientation towards customer service. Though there were formal procedures in the company service charter on how to deal with customer complaints, statistics on response times to complaints were not easily available, and the use of available data for performance management was poor. Furthermore, benchmarking was in its infant stage. It focused on tariffs, water quality, level of services and efficiency. However, neither the utility managers nor the Board used such data for decision making. Finally, although the company introduced a service charter, it did not carry out any market survey to detect the perceived customer satisfaction.

In synthesis, the company suffered from the following problems:

- Negative operating income flows were regularly covered by the Municipal administration;
- Lack of coordination among managers was a cause of lack of awareness of how each sector was contributing to the organization's results;
- The management information system was characterized by the production of a number of reports that were mainly responding to bureaucratic routines, instead of strategic information needs;
- Performance evaluation was perceived as a "*weapon*", through which managers could have been blamed for bad performance, rather than as a tool to enhance efficiency and effectiveness.

With the aim to create a shared vision of business strategy, to foster a deep cultural change, to stimulate communication among managers, to increase competitiveness and avoid strategy disconnections among the different levels of the organization, a "dynamic" balanced scorecard (DBSC) was designed (Bianchi and Montemaggiore 2008). This project involved the company key managers in modeling end-results, performance drivers and strategic resources, and framing the system structure behind the company's performance.

The Board organized a number of meetings with top and middle managers with the purpose of designing an information system that could be used to monitor performance. The final result was a long list of activity indicators, included in a 40-page report. Neither a common strategy was designed, nor were causal linkages connecting these activity measures provided.

With the aim of translating the produced list of indicators into a BSC map, the project team conducted several interviews with AMAP's key managers. These interviews fostered the elicitation of their tacit knowledge about business processes and causal relationships between policy levers, performance drivers (lead indicators), and end-results (lag indicators). This allowed the project team to reduce significantly the long list of measures included in the initial report.

The proposed strategy consisted mainly in improving the company image by promoting a higher efficiency and effectiveness in the provision of service. Such a goal could have been achieved by increasing the availability of water sources (see the driver "Refinement %" and the end-result "Pumped water/day" in Fig. 4.1) and hence, the volume of water distributed to households (end-result). In this regard, AMAP was able to satisfy only 60 % of the standard consumption *per capita* stated in the service charter. Therefore, an increase in the volume of distributed water would have improved customer satisfaction, income, and cash flows (end-results), through higher revenues and lower unit costs (since a larger volume of supplied water would have reduced the overheads and the fixed costs per cubic meter).

The improvement in customer satisfaction (by a better service) and of shareholder satisfaction (by higher financial results) would have enhanced company image.

For this reason, a great deal of effort was devoted to the search for new sources and to the acquisition of the right to exploit a larger percentage of the existing sources. With this purpose in mind, AMAP evaluated the opportunity to invest in the construction of a purification plant for the treatment of wastewater. Basically, in

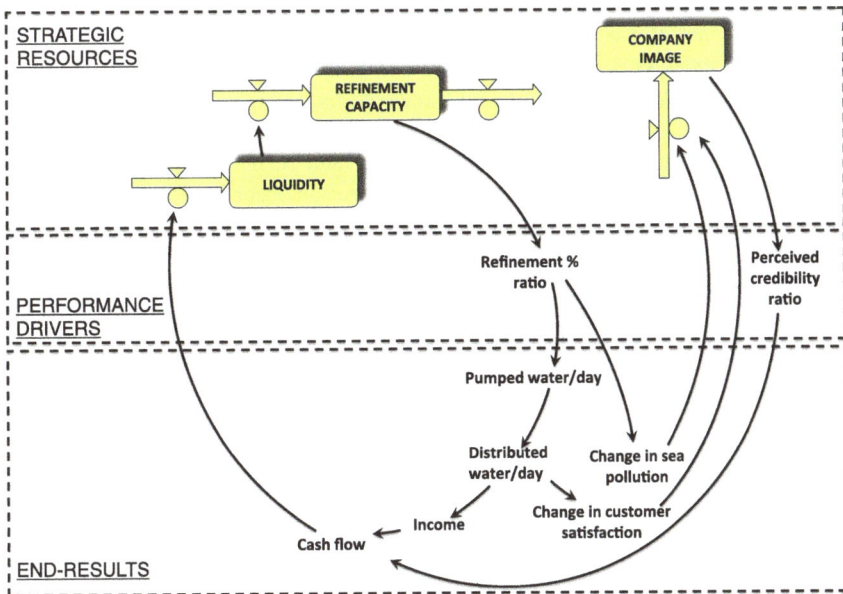

Fig. 4.1 DPM chart showing reinforcing loops enhanced by refinement investment policy

the purification plant, the sewage collected is subjected to a specific purification process so that it can be used for agricultural purposes. Because of this investment, the company could have distributed the purified wastewater to farmers, thereby increasing the volume of drinkable water distributed to households. Furthermore, according to the company's management, purification of the wastewater would have improved sea pollution conditions.

The planned investment, therefore, would have given evidence of the company's commitment to the cleanness of the seashore and, hence, to the improvement of the community's life quality. This, in turn, would have further enhanced the company's image and, therefore, its advantage over other potential competitors in the management of integrated water services in the area. Image affects the perceived credibility of the firm in the financial market and towards the local government (performance driver). Therefore, an improvement in company image may allow the utility to better negotiate funds to borrow from different stakeholders, and therefore to increase cash flows to reinvest in refinement capacity.

Decision makers were conscious that the purification policy would have led to a greater volume of distributed water and to lower sea pollution and, hence, to higher customer and community satisfaction and, eventually, company image. Nevertheless, they were still evaluating the adoption of this policy because of the high investment and production costs, which would have had a negative impact on company financial results, reducing shareholders' satisfaction and, consequently, the image of the company as an efficient administrator of the municipal water service.

The DPM chart in Fig. 4.1 shows the commented feedback relationships hidden in the company vision.

However, a problem AMAP also had to face was the high leaking rate of its pipelines. In fact, it was necessary to improve the quality of pipelines by replacing the quite old existing distribution network. The obsolescence of pipelines caused a high rate of leakage, which significantly reduced the volume of water distributed to households.

If the volume of pumped water grows faster and stronger than the distribution capacity, an increase in distribution capacity utilization occurs and, consequently, the leaking rate increases. In fact, the more water is pumped through the pipelines, the higher is the pressure, and the bigger is the volume of leakage through the holes, joints, etc. As a consequence of this, the cost per cubic meter of distributed water increases.

Therefore, the described phenomenon, on the one hand, contributed to customer dissatisfaction, and on the other hand further worsened the efficiency of the distribution process and, hence, the company's financial results.

The balancing loop "B1" illustrated in Fig. 4.2 (which is a simplified version of a portion of the stock-and flow model underlying the DBSC) frames the potential diminishing returns of an investment policy that is only oriented to improve the distribution capacity, without considering the quality of the distribution network. The figure also shows that balancing refinement capacity and distribution capacity investments may allow the firm to pursue a sustainable growth.

As illustrated in Fig. 4.3, the obsolescence of pipelines was modeled as a driver ("Pipeline quality ratio") that can be affected by investments in distribution

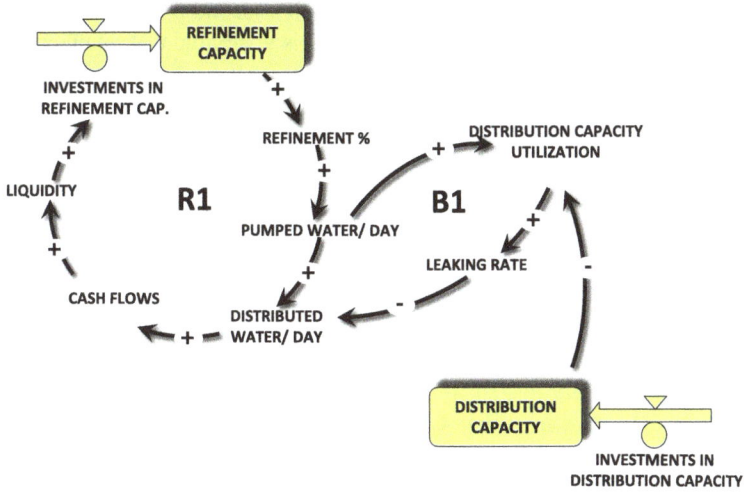

Fig. 4.2 Balancing loop acting as a limit to the growth in distributed water per day

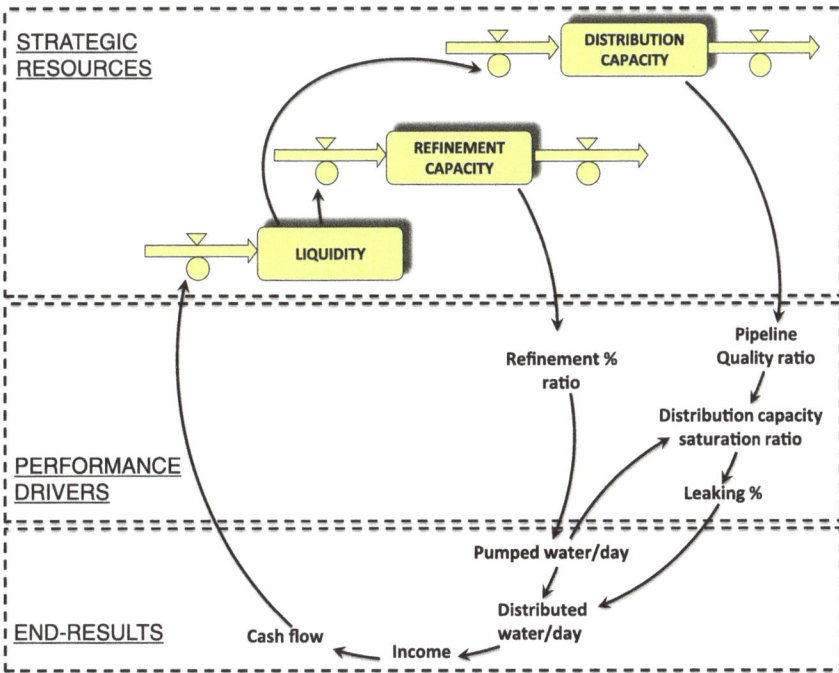

Fig. 4.3 DPM chart showing effects on performance from a combined refinement and distribution capacity investment policy

capacity. This driver affects the leaking % through a second performance driver, i.e. the "Distribution capacity saturation ratio". This is a variable that is affected by the obsolescence and capacity of the distribution network, and by its usage (i.e. the pumped water/day). The more water is pumped, the higher capacity saturation will be. The more distribution capacity investments are done, the lower—other conditions being equal—the distribution network saturation will be. The leaking % has been modeled as a normalized graph function where a higher distribution network saturation will determine a non-linear increase in the leaking rate. The leaking % will reduce the volume of distributed water/day, which would be possible through a given volume of pumped water/day.

Another reinforcing loop associated with distribution capacity investment policy is related to the allocation of more auxiliary workers to the repair of breakdowns in the distribution capacity system (end-result), caused by its obsolescence (performance driver). As shown in Fig. 4.4, the higher the number of auxiliary workers allocated to repairing tasks, the shorter the time to fix breakdowns will be (performance driver). This will increase service and customer satisfaction (end-result), leading to higher company image and capability to negotiate funds to boost cash flow (end-result). Higher liquidity resulting from increased cash flows could be reinvested in hiring more auxiliary workers. A larger auxiliary worker staff will allow the firm to further reduce the time to fix breakdowns.

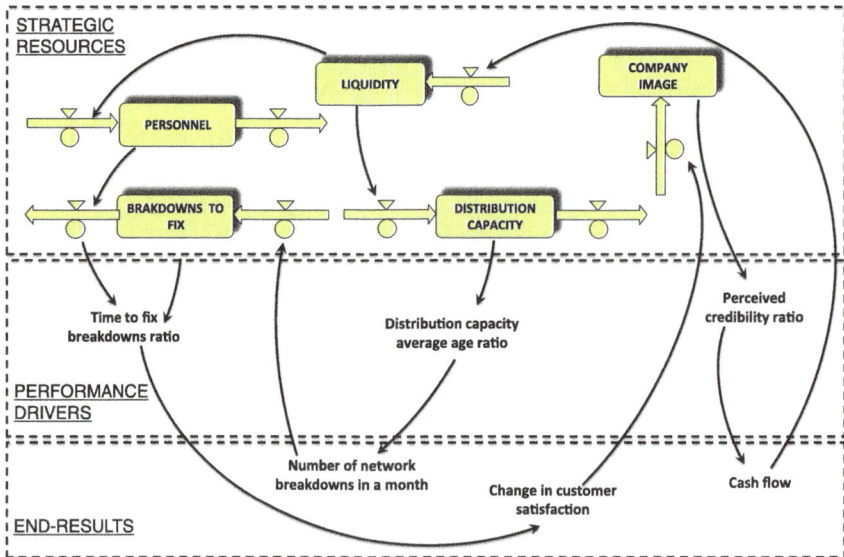

Fig. 4.4 DPM chart showing effects on performance from allocation of auxiliary workers to promptly repair network breakdowns

The stock-and-flow model was developed around four sectors,[2] i.e.:

1. Distribution, which analyzes the supply and distribution process of the water and the aging process of pipelines;
2. Sewer, which refers to the collection and purification of wastewater;
3. Human resources, which describes the allocation of the auxiliary workers between the maintenance activity and the activity of service suspension to overdue clients;
4. Financial sector, where the dynamics of the net income, cash flows and other financial resources are framed.

Different causal linkages exist between the four sectors. In the financial sector, resources available for investments in distribution and sewer capacity are calculated, based on liquidity and the possibility of borrowing funds. First, the model proportionally allocates such resources to replace old distribution pipelines and decayed purification capacity. Then, the residual resources are proportionally allocated to increase both pipeline and purification capacities.

In turn, the renewal of pipelines and increase in purification capacity produce costs and debts, which affect the company income, loans and liquidity.

Moreover, the "breakdowns" flow in the distribution sector determines the maintenance backlog that the auxiliary workers have to reduce. The maintenance activity generates an increase in the fixed pipeline capacity, which enhances the distribution capacity, and costs, which reduce the company income.

A business simulator, embodying both the SD and the accounting models portraying financial statements, was built on the basis of the above-mentioned sectors. By using this tool, managers can easily:

- Input the initial model parameters according to company data;
- Insert the company objectives in the different perspectives through a BSC chart;
- Experiment with different policies under various scenarios through a control panel including policy levers and a scenario-setting board;
- Evaluate company strategy through several tables and graphs, reporting the simulated impact of the interrelated set of policies according to the selected performance indicators.

From the case discussed in this section it is possible to appreciate the benefits of DPM to support strategy design and planning in public utilities, where a deep cultural change and major performance improvement are strongly required.

In the AMAP case, strategic mapping and simulation through the SD methodology has proved to successfully enhance managers' learning and capability to identify causal relationships between policy levers and company performance, and better communicate strategy with stakeholders.

---

[2]More details on the structure of the model are in Bianchi and Montemaggiore (2008).

4.3 Applying Dynamic Performance Management to Public Utilities (Continued): Garbage Collection at the City of Winston-Salem

The city of Winston-Salem is located in North Carolina (USA), with a current population of about 200,000. With the goal to foster service delivery, the city started performance measurement in the '80s. With the same purpose it started the North Carolina Benchmarking Project (Rivenbark 2001) in 1995, in collaboration with the UNC—School of Government, to compare the city's performance and cost data against other municipalities in North Carolina.

Data were compared from seven municipalities in the state across different service areas (residential refuse collection, household recycling, yard waste, police patrol, police investigations, emergency communications, and street maintenance).

The first report showed a very low productivity in tons collected per full-time equivalent (FTE) in residential refuse collection. One explanation of this result is that Winston-Salem was one of three municipalities participating in the benchmarking project that used backyard garbage pickup rather than curbside collection. However, when Winston-Salem compared itself only to backdoor providers, the city was still behind in tons collected per FTE (Ammons 2000).

During the FY 1997–1998 annual budget process, staff members from the budget department and the sanitation division met with the city manager to discuss options for reducing this performance gap in the service area of residential refuse collection. They decided not to renew a private contract for garbage collection in an annexed area of the city and to adjust current city routes to provide coverage within existing resources. This policy increased garbage collection productivity by approximately 30 % (Rivenbark et al. 2005). However, the performance and cost data report for FY 2000–2001 showed that Winston-Salem remained inefficient when compared with the other participating municipalities in the service area of residential refuse collection. A possible explanation of this performance gap is that several of the other municipalities had embraced technology to productivity (Rivenbark and Pizzarella 2002).

Based on the benchmarking data, in 2004 the city council decided to approve voluntary curbside collection, with approximately 15,500 households agreeing to participate in the new program.

In spite of this new policy, the city's residential refuse collection program was still inefficient, with respect to the benchmarks.

The city finally made the decision to mandate curbside collection in October 2010. *What prevented the city from making this transition in 2004 that would have produced an annual savings of $1,800,000 (i.e. 22.5 % of the budget deficit)?* One possibility is that an understanding of the complete system was not shared among the involved stakeholders, which prevented an effective use of available performance data to drive change.

A conceptual DPM model was sketched in order to illustrate the hidden feedback structure underlying the service delivery system's performance (Bianchi and Rivenbark 2014).

A major difficulty in developing such model has been the lack of explicit data that can support the calibration of possible system structures that might explain recorded performance data. We started by modeling population and collection points of the city's residential refuse collection system, where patterns over the period 1997–2010 were well defined by available benchmarking records.

The decision to start the analysis from such variables was also associated with the fact that an increase in the collection points determines an increase in the workload and—other conditions being equal—an increase in collection costs. However, a decrease in the quality of garbage collection services or an increase in taxation—due to higher garbage collection costs—may dampen the city's attractiveness, and therefore may contribute to increased (though with a delay) population outflows as shown in Fig. 4.5.

Both population and collection points were identified as strategic resources in the system because they correspond to factors that primarily affect the performance of the sanitation division and the city as a whole. The dynamics of collection points were modeled through co-flows depending on population accumulation and depletion rates and on the "average household size". This parameter was estimated

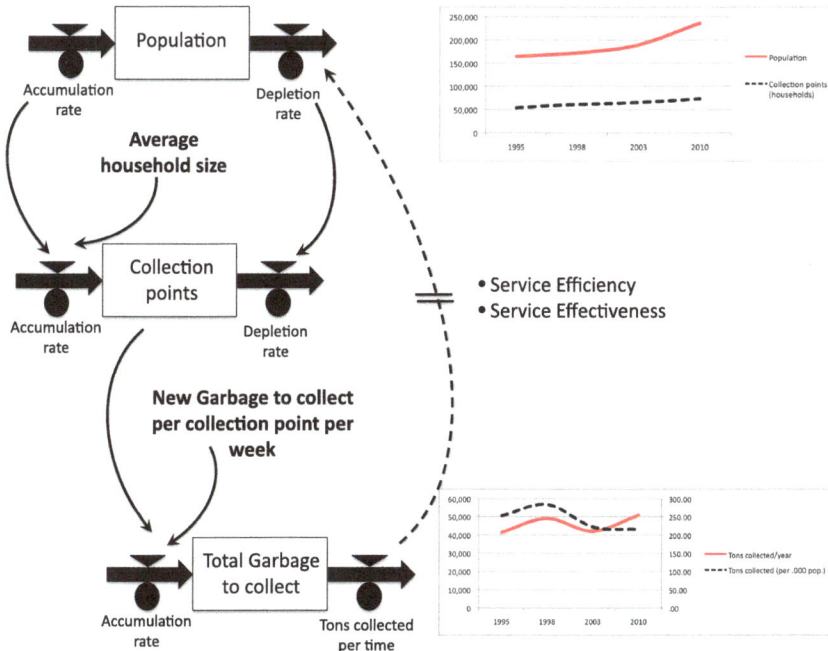

Fig. 4.5 Mapping population and collection points

based on a ratio between population and collection points using historical data. The upper section of Fig. 4.5 highlights how an outcome perspective should characterize the sanitation division's policies.

In planning capacity and service provision, setting performance standards, and evaluating results, the division should interact with other divisions in the same municipality with a view to understanding different factors impacting workload (total garbage to collect) and the driving forces affecting population flows. Relevant to this point—though not explicitly linked to sanitation services—are policies regarding: housing, infrastructure, education, healthcare, enterprise services, and public safety.

In order to model the effects of the sanitation division's performance on population, we will explore now the factors impacting the outflow depleting the "total garbage to collect" stock (tons collected per time). Such factors will affect service efficiency (municipal financial needs) and effectiveness (the perceived quality of provided service).

Figure 4.6 shows how tons collected per week should be tracked as an end-result that is affected by the performance driver "tons collected per working hour" and by the "working hours per week" (capacity resource).

The variable "tons collected per working hour" gauges worker productivity, which is in turn affected by the level of investments in service automation (strategic resource) and by the backyard collection ratio (performance driver). This is the percentage of collected garbage from backyards of the total (ratio between two

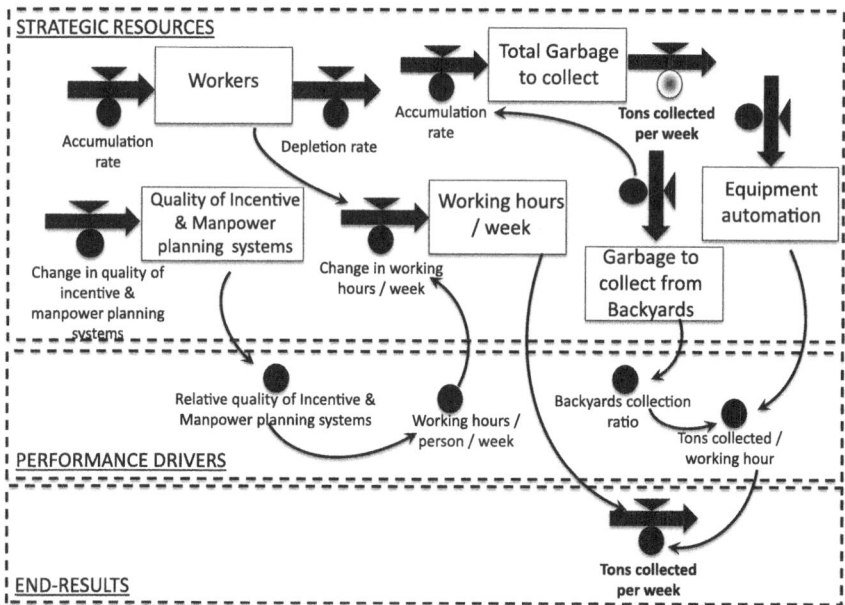

Fig. 4.6 Modeling worker productivity

stocks). As the ratio increases, worker productivity decreases because fewer tons will be collected per working hour and fewer tons of garbage will be collected each week unless more staff is hired.

Another performance driver, which indirectly affects the tons collected per week, can be referred to as the quality and consistency of incentive planning systems. This variable is affected by investments into an intangible strategic resource stock. The design of better human resource planning and incentive systems will be likely to contribute to increased individual productivity.

A higher individual productivity would contribute (given a stock of available workers) to increase the actual total working hours per week, which would increase the tons of garbage collected per week. Therefore, the same result could be achieved through a multitude of combined policies.

In this regard, DPM simulation models can support scenario analysis.

In order to track the perceived quality of provided service (effectiveness), the perceived time to serve households was considered as a strategic success factor. This variable can be modeled as a ratio between the "total garbage to collect" and its outflow, as shown in Fig. 4.7.

A change in such ratio does not immediately affect the municipality's reputation toward its service users (perceived time to serve households). In fact, an information delay smoothes changes in the actual time to serve households because it may also take several weeks by citizens to change their mindset (strategic resource generated by management routines) about the provided service.

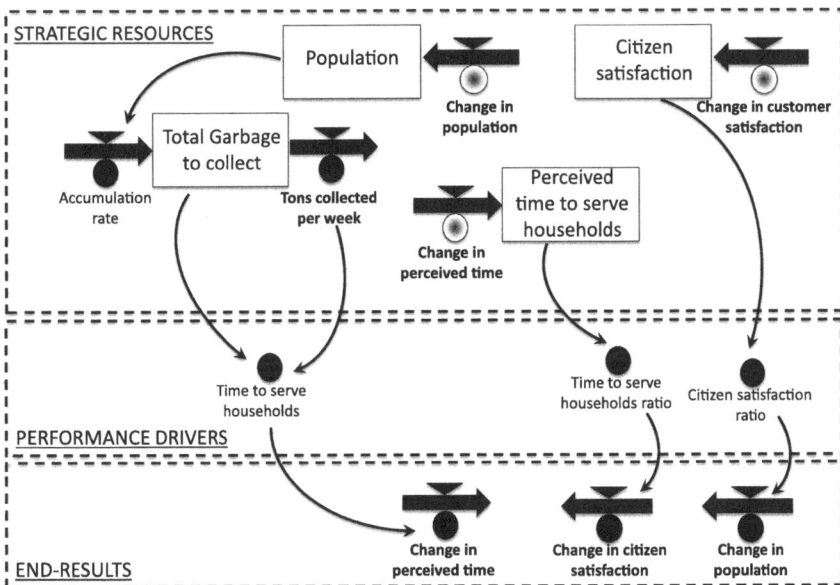

Fig. 4.7 Modeling service levels

A change in such mindset will contribute to change in the long run the overall citizen satisfaction rate (outcome end-result), which will change the population stock. The change in population (which is also an outcome end-result) is due to the hypothesis that efficiency and effectiveness in residential refuse collection would contribute to determine—in the long run—a decision by more citizens to move to and from neighbor municipalities.

This case illustrates how DPM can be used to help local officials move from the adoption of performance measurement to performance management.

In applying DPM to the service area of residential refuse collection system for the city of Winston-Salem several benefits of using this methodology were identified that can play an important role in improving the quality of organizational performance management systems and that can help local officials use performance data for making critical policy decisions.

They include framing trade-offs in time and space associated with alternative scenarios, understanding how different policy levers impact the accumulation and depletion of strategic resources, and determining how performance drivers affect end-results. DPM also can help local officials with establishing goals and objectives, linking the political to the managerial level, and focusing attention on selecting relevant targets and evaluating results.

The evaluation process is critical, where DPM is used to capture intangibles, system delays, and nonlinear relationships and is used to avoid common errors of causalities.

4.4   Applying Dynamic Performance Management to Municipalities: Linking Strategic Goals and Departmental Objectives Through "Dynamic" Balanced Scorecards

Between the end of the '90s and the beginning of the 2000s, the town of Hillsborough (with a population of about 6000), had been advancing its performance measurement system. Departments were encouraged to develop and use performance measures in planning and decision-making. Incremental budgeting was replaced by a planning process embracing program budgeting, and a two years financial forecasting, aimed to anticipate future problems and needs (Rivenbark and Peterson 2008).

The strategic priorities addressed quality of life, growth management, economic development, superior services, and community safety.

However, the town needed an approach to link strategic goals and departmental objectives. In November 2006, the board adopted the balanced scorecard (BSC) as a methodology that could support these connections.

The board used the strategic priorities to guide the development of town objectives (Fig. 4.8). The city manager encouraged each department to develop a

PERSPECTIVES   TOWN OBJECTIVES

Customer SERVE THE COMMUNITY	Strengthen Citizen Involvement & Access	Preserve Cultural & Natural Resources	Reduce Crime & Increase Citizen Safety	Enhance Community Sustainability	Expand Recreation, Walkability, & Connectivity	Improve Satisfaction with Services
Internal Business Process RUN THE OPERATIONS	Enhance Emergency Preparedness	Provide Responsive & Consistent Services	Improve Communication & Collaboration	Excel at Staff & Logistical Support		
Financial MANAGE RESOURCES	Maintain Fiscal Strength	Invest in Infrastructure	Develop Long- Term Financial Plans	Deliver Efficient Services		
Learning & Growth DEVELOP PERSONNEL	Develop a Skilled & Diverse Workforce	Support Training, Learning, & Growth	Enhance Relations with Other Entities			

Fig. 4.8 Performance perspectives and town's strategic objectives (from: Rivenbark and Peterson 2008, p. 34)

robust set of performance measures that could allow the organization to understand how departmental services ultimately affected the attainment of the strategic priorities, mission, and vision of the organization. This approach aimed to combine workload and outcome measures.

As shown in Fig. 4.9, it was found that the performance of an internal service function such as motor pool fleet maintenance (which provides services to other town departments) might affect town objectives in each of the four dimensions. Furthermore, each motor pool fleet maintenance initiative may contribute to align different town's strategic objectives. They may range from those the department can directly affect through its outputs (end-results), to those that it can affect through its outcome end-results.

For instance, the initiative defined as "Emphasize completion of Automotive Service Excellence (ASE) certification program" primarily aims to attain a given target workload of ASE certifications. Such output measure affects the town's strategic objective named "Develop Skilled and Diverse Workforce", that pertains to the BSC "Learning and Growth" dimension.

By improving the employees' skill level, through the completion of ASE programs, the department may also increase the percentage of services performed on schedule; such measure is linked to the "Manage efficient and effective preventive maintenance programs" initiative, that aims to contribute to the "Excel at staff and logistical support" town's strategic objective (which pertains to the BSC "internal processes" dimension).

If we frame the described causal relationships through the lenses of DPM, we may observe that the percentage of services performed on schedule is a performance index, rather than a measure driving the output end-result that could be referred to the number of maintained vehicles in a give time span. Such performance index cannot be used efficiently as a surrogate measure of the average time to manage preventive maintenance, which is a performance driver of the vehicles failure rate

Perspective	Townwide Objective	Motor Pool Initiatives "What will the department do?"	Measures "How will the department know when the desired results are being achieved?"	Performance Data	
				Prior-Year Actual	Target
Serve Community	Preserve cultural and natural resources	Prevent contamination of stormwater by capturing and recycling used oil and antifreeze	Amount of used motor oil and antifreeze recycled	501 gallons	500 gallons
	Enhance community sustainability	Buy parts and supplies locally when cost-effective, thereby supporting local businesses	Percentage of parts and supplies purchased locally	59%	60%
Run Operations	Excel at staff and logistical support	Manage efficient and effective preventative maintenance program	Percentage of services performed on schedule	88%	90%
Manage Resources	Deliver efficient services	Provide efficient services by preparing monthly report listing repair and service cost per vehicle	Average maintenance cost per vehicle	$855	$850
Develop Employees	Develop skilled and diverse workforce	Emphasize completion of Automotive Service Excellence (ASE) certification program	ASE certifications obtained	11	24

Above the table, circled: "Reduce Crime & Increase Citizen Safety"

Fig. 4.9 Aligning departmental measure/objectives with town's strategic targets: motor pool fleet maintenance (adapted from: Rivenbark and Peterson 2008, p. 35)

(end-result). Such index cannot be used as a surrogate of the average maintenance time since it has a dimensionless unit of measure. On the contrary, the time to manage maintenance is a ratio between the backlog of cars under preventive maintenance and the outflow of cars that have been maintained in a given time span, which is also computed in the numerator of the index.

Figure 4.10a, b frame the effects of employee training policies on the department performance, in terms of output and outcome, respectively. The first DPM chart illustrates how the employees' skill level affects (together with other strategic resources, such as staff and equipment) available capacity, which is measured in terms of "productive" working hours. So the potential (or nominal) capacity is related to available staff and equipment. Such capacity may either increase or decrease, depending upon the staff skill level, which determines the productive (or available) capacity. If such capacity is lower than the benchmark (i.e. desired capacity), the preventive maintenance capacity ratio (performance driver) will be lower than 1. This will imply a lower efficiency in performing maintenance, in respect to the normal level. Such effect is modeled through a graph multiplier that is called "Effect on time to perform a task" (see Fig. 4.10a).

Figure 4.10b describes the time to manage preventive maintenance ratio (i.e. the fraction "actual/desired" maintenance time) as a driver affecting the vehicles failure rate. Such driver is related to the average time it takes for the department to deplete

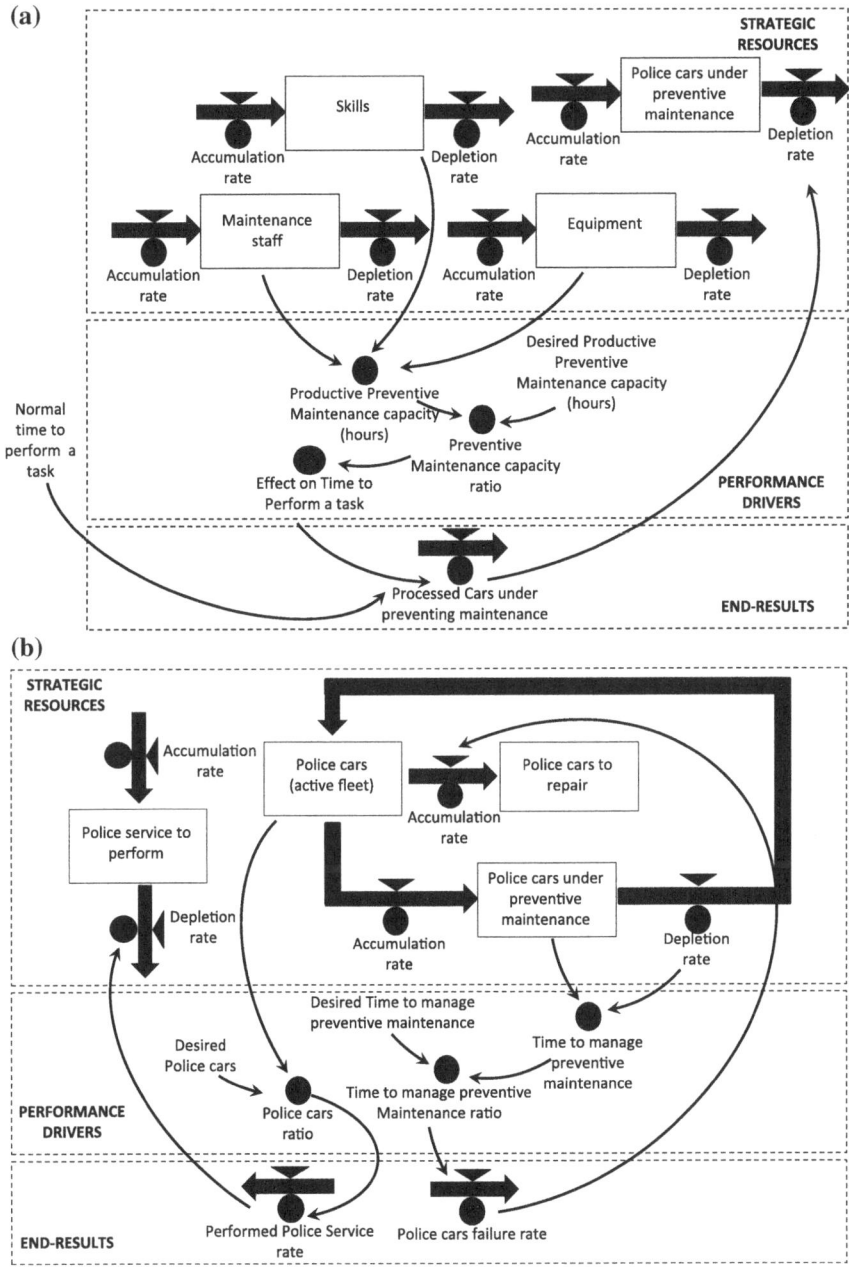

Fig. 4.10 a Effect of employees' skill building policies on the capacity ratio, time to perform a task and number of processed vehicles under preventing maintenance (output measure). **b** Effect of time to manage preventive maintenance on vehicles failure rate and quality of community service (outcome measures)

its backlog of cars under preventive maintenance. This variable is an effect of the maintenance activity's end-result that was previously described as affected by available capacity. In fact, it is calculated as a ratio between the stock of cars under preventive maintenance in a given time and the outflow of cars that have received preventive maintenance in given time span.

If we apply this analysis to the internal maintenance services performed by such department to the benefit of the police department, a time to manage preventive performance that is higher than the benchmark would imply a risk of a higher police cars failure rate, in respect to the normal. This is an outcome performance measure to which the motor pool fleet maintenance department could be made accountable.

The effect of such measure might be useful to shed light also on the indirect contribution of the motor pool fleet maintenance department to the city strategic target "Reduce Crime and Increase Citizen Safety". As Fig. 4.10b shows, the effect of an increasing volume of police car failures over a time span would reduce the active fleet stock (strategic resource). This would reduce—other conditions being equal—the police car ratio, which is a driver of a decreasing performed police service rate. A lower service rate would giver rise (other conditions being equal) to an increasing police service backlog to perform, which might mean a risk of an increasing crime and a decreasing citizen safety.

Figure 4.11 frames the trade-off in time related to employee skill building policies. An intensive training policy is able to increase average training time per staff, which will drive in the long run a change in skills (outcome end-result).

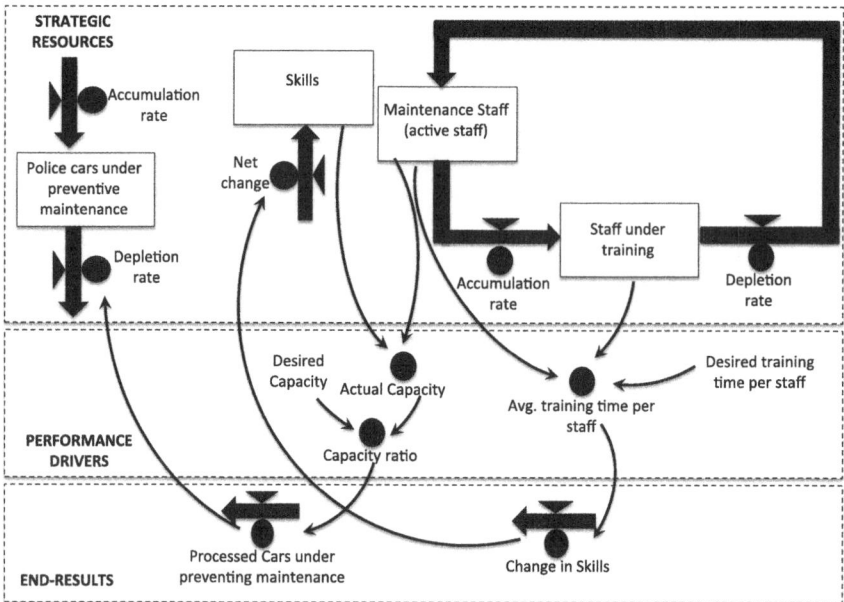

Fig. 4.11 Trade-offs associated with employees' skill building policies and related drivers, output and outcome end-results in the short and long run

However, in the short run this policy would decrease (other conditions being equal) the capacity ratio, which would lead to a lower productivity, in terms of processed cars under preventive maintenance (output end-result).

DPM, particularly if enhanced by computer simulation, can support decision makers to frame these trade-offs and perform scenario analysis that may enhance a feed-forward approach to planning and control.

4.5   Using Dynamic Performance Management to Overcome a Myopic View in Designing Performance Measurement Systems in Municipalities: The Case of Policing

This section will illustrate the potential benefits of applying DPM to the design of performance measurement/management systems in the public sector to prevent, detect, and counteract behavioral distortions caused by improper setting of performance measures. A number of exemplary cases of municipal policing will be discussed.

A problem that is often encountered when social policy programs encompassing different interconnected sectors (such as immigration, education, healthcare, safety, welfare, and housing) are designed and implemented is the focus of each player on a limited set of output targets that are to be achieved without building on an understanding of the process and factors that lead to them and generate outcomes in the long run.

Behn (2011) has pointed out the difficulties and the risks hidden in the adoption of performance measures as part of reward and incentive frameworks in the implementation of social policy programs. Concerning this, he refers to the "Campbell's Law" i.e.: "The more any quantitative social indicator is used for social decision-making, the more subject it will be to corruption pressures, and the more apt it will be to distort and corrupt the social pressures it is intended to monitor" (Campbell 1969).

In this context, an inconsistent design of performance measures may generate organizational behavioral distortions, whose dysfunctional implications imply a risk of inverting means with ends, or of focusing on only a limited set of short term targets, regardless the long term effects that the undertaken policies will generate in the larger relevant system.

To describe the dysfunctional effects of performance management where phenomena are characterized by a weak capability of performance indicators to measure and affect performance, the term "performance paradox" has been used by the literature (Meyer and Gupta 1994; Van Dooren et al. 2010, p. 165)

Van Thiel and Leeuw (2002, p. 271) discuss the unintended effects generated by the use of an aggregate performance measure of the percentage or number of crimes solved by the Dutch police. If one focuses on only the percentage of crimes solved

Fig. 4.12 Unintended effects
produced by the use of an
aggregate performance
measure by the Dutch police

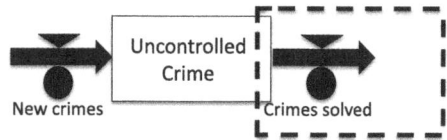

as a performance measure, a decrease in it might suggest that police performance is
deteriorating. However, the reduction in such measure could result from an increase
in the stock of crimes because of an increase in the volume of new crimes. The
absolute number of crimes solved might remain constant or even increase, yet the
ratio of inflow (new crimes) to outflow (solved crimes) could also increase.
Likewise, an increase in the crimes solved over a number of years should normally
lead to a reduction in the stock of criminals; thus making it more difficult to
continue improving or even to keep the absolute number of crimes solved.
However, a reduction in this measure resulting from a drop in the stock of criminals
would not indicate performance deterioration (Fig. 4.12).

The DPM chart illustrated in Fig. 4.13 (Bianchi and Williams 2015, p. 409)
shows how police deploy staff and equipment (strategic resources) to address
inflows of crimes and criminals, converting them to outflows through solved crimes

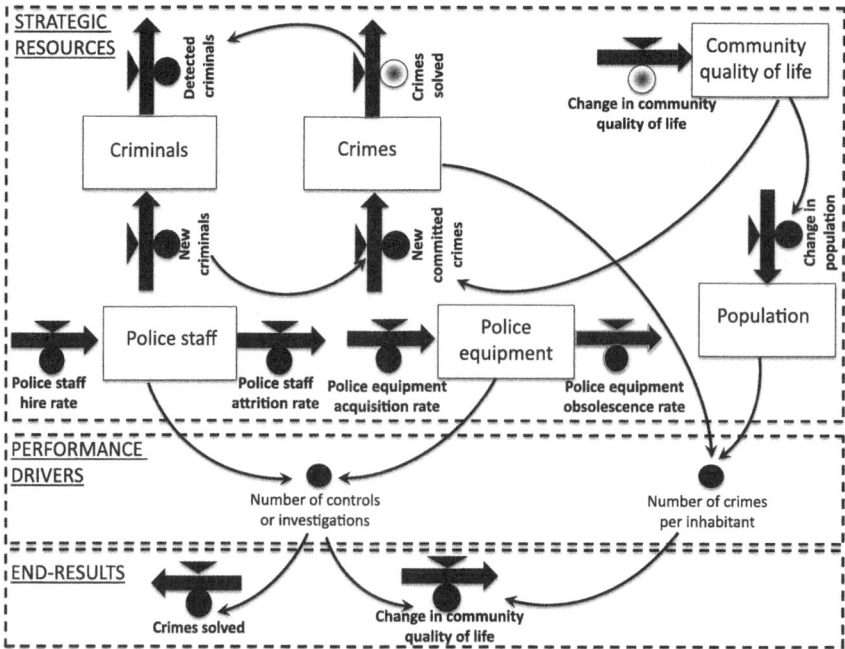

Fig. 4.13 Dutch policing DPM chart illustrating the unintended effects on performance outcomes
from a focus on a single performance measure

and identified criminals. The purpose of these actions is to investigate and control crime (performance driver) and thereby solve crimes (output end-result measure).

To counteract the problem of diminishing returns, the police might increase the pressure on the community in search of a number of crimes to solve, eventually reaching an intolerable level of intrusion into the lives of community members.

So in this example, the risk of gaming is associated with the possibility that without a deliberate purpose, the police staff and other strategic resources might be deployed to pursue performance drivers, such as the number of controls or investigations undertaken, with the intent to achieve numerical targets.

If solving crimes and reducing the corresponding stock of criminals affects the community's quality of life, an excess of controls and investigations may offset in the long run the positive effects of the described policy, as shown by the second effect on the right side of the end-results section of Fig. 4.13. This might reduce the community's quality of life (outcome measure). It could also generate both a reduction in the population level and an increase in the inflow of new crimes. Therefore, an excess of focus on a single policy might generate specific dysfunctional effects over time. In this case, such effects could be related to an increase of the number of crimes per inhabitant (performance driver), which would further reduce the community's quality of life, a vicious reinforcing loop.

By extending the boundaries of the observed system, one may reduce the risk of performance paradox. For example, rather than focusing on only the crimes solved, the police might also be made accountable on a set of outcome measures that would describe the effectiveness of their crime prevention efforts. This would allow the police to take an active role in designing and implementing broader social policies, in coordination with others, such as schools, social services, nonprofits, and associated stakeholders.

Figure 4.14 shows how DPM can help decision makers to extend the system's boundaries. It shows that, when considering additional end-results related to the stock of criminals, two more flows should be taken into account: the new criminals and the new potential criminals. Correspondingly, the stock of potential criminals is added as a relevant strategic resource to include when formulating policies that may contribute to sustainable community security. In addition, a second set of performance drivers is included in this alternative context: the number of preventive actions.

Other problematic behavioral distortions emerging from the "CompStat" program and other similar programs adopted over the last two decades by different public sector organizations around the world will be now framed through the lenses of DPM.

CompStat is a statistics-based performance measurement system that the New York Police Department adopted in 1994 to motivate employees, with a specific focus on precinct commanders, and to sharply reduce crime (Behn 2003, p. 591). Despite reports of success in fighting crime, CompStat has been mentioned as a cause of dysfunctional behavior and performance paradox.

The Campbell's Law generates its own effects particularly in social contexts as crime control: "False arrests have been identified as the result of arrest quotas…

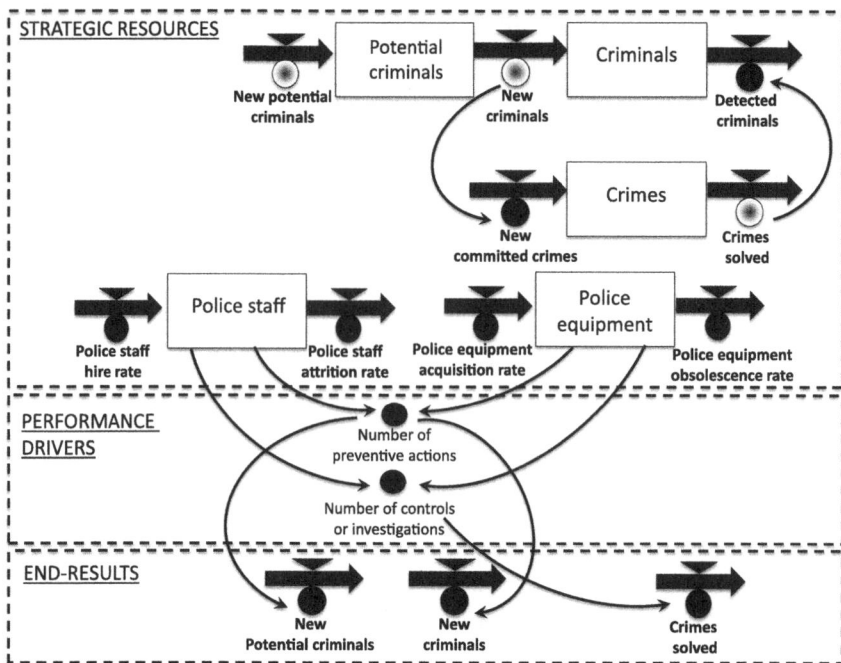

Fig. 4.14 Extended Dutch policing DPM chart

arrest quotas may encourage police to focus on less difficult and important arrests at the expense of more significant and arduous arrests." (Eterno and Silverman 2012, p. 11). Manipulating or "fudging" crime data has been said as a recurring practice adopted by the police to generate the expected numbers to report; for instance: "misclassify crimes from felonies to misdemeanors, under-value the property lost to crime so it's not a felony" (Eterno and Silverman 2012, p. 27).

Figure 4.15 illustrates the distorted effects on reported crime produced by performance measures and by a reward system focused on only short-term and "easy to achieve" output measures.

The balancing loop "B1" would be the functional response of the performance management system to the actual crime level. An increase of this level would imply a higher effort to suppress crime, leading (after a delay) to a reduction of crime. However, the use of an unbalanced set of performance measures and of short-term "output-only-based" reward mechanisms could hamper the dominance of the described balancing loop by fostering two other loops: the reinforcing vicious loop "R" and the balancing loop "B2".

The reinforcing loop "R" is generated by an increasing effort to solve "easy" crimes (due to the existing set of performance measurement and rewards systems) and, therefore, to report an increasing pattern of such kinds of crimes. This generates an increase in the distributed rewards, which in turn gives rise to a "search"

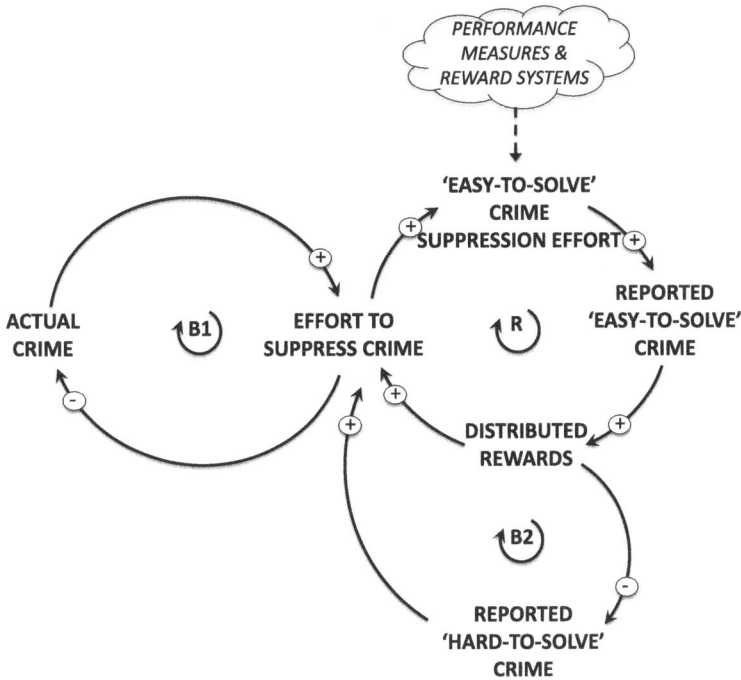

Fig. 4.15 Distorted effects of short-term output measures

for new "easy-to-solve" crime to deal with and report. Though this vicious loop can contribute to reduce crime (in absolute terms), it might not generate the desired effects in dealing with "hard-to-solve" crimes. On the other hand, the described system could also affect the statistics regarding such crimes through data fudging. The distribution of rewards may lead the police to reduce reported hard-to-solve crime and shift part of it to the lower rank of "easy-to-solve" crime. This may generate a further effort towards the suppression of easy-to-solve crimes. So while the loop "B2" dominance replaces the loop "B1", it is also the basis for the loop "R" to further and progressively generate its dysfunctional effects.

The DPM chart illustrated in Fig. 4.16 frames the dysfunctional effects caused by lack of alignment between performance management and rewards systems. When such systems generate "false reporting" on "easy-to-solve" cases, performance management becomes a means for the organization to reach the desired rewards. In the described context, the performance management system takes an ancillary role in respect to the rewards system. The use of strategic resources is primarily diverted by the rewards system to the execution of controls and preventive actions on "easy cases". This effort contributes to meet the output targets, shown as flows of "easy cases" solved.

The described behavior generates two main effects. First, it allows the organization to distribute rewards based on the achievement of the target number of cases

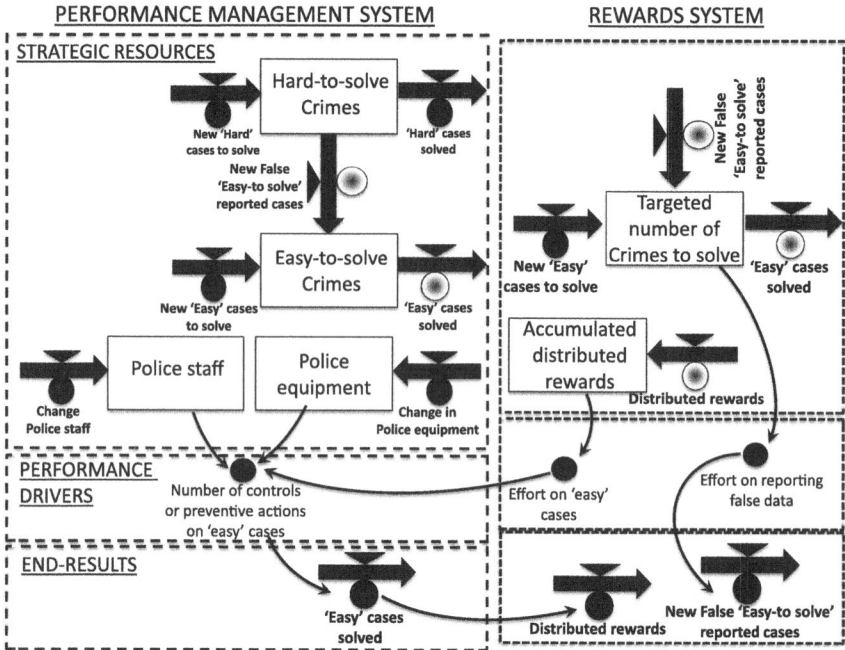

Fig. 4.16 Detecting distortions in rewards and performance management systems through DPM

to solve. The large volume of solved "easy cases" reinforces the process and leads to an expectation of more similar cases, which implies that organization redirects efforts to such cases. Second, although an increase in the outflow of "easy cases" solved should lead to a reduction in the stock of "easy-to-solve" crimes, this does not happen, because of the inflow of new false cases. This variable is shown as an output of the rewards system at the bottom right of Fig. 4.16. The volume of false cases is proportional to the effort on reporting false data, which is proportional to the stock of targeted number of crimes to solve (Bianchi and Williams 2015, p. 415).

Therefore, an abnormal behavior of the rewards system here generates a set of end-results, which may allow the distribution of rewards regardless of organizational sustainable performance. A further vicious effect can be that unaddressed serious crimes might allow an increase in hard-crime criminals, who become skilled, leading to a gradual long-run reduction in the city's quality of life.

The insights emerging from the previous examples suggest measures to counteract these problems. Figure 4.17 shows that improving the performance management system may lead to greater reporting promptness and selectiveness. This implies investing in performance measurement and information systems and also in organizational design and human capital.

For instance, as shown in Fig. 4.15, improving the promptness and selectiveness of reporting may allow decision-makers to design policies that better balance

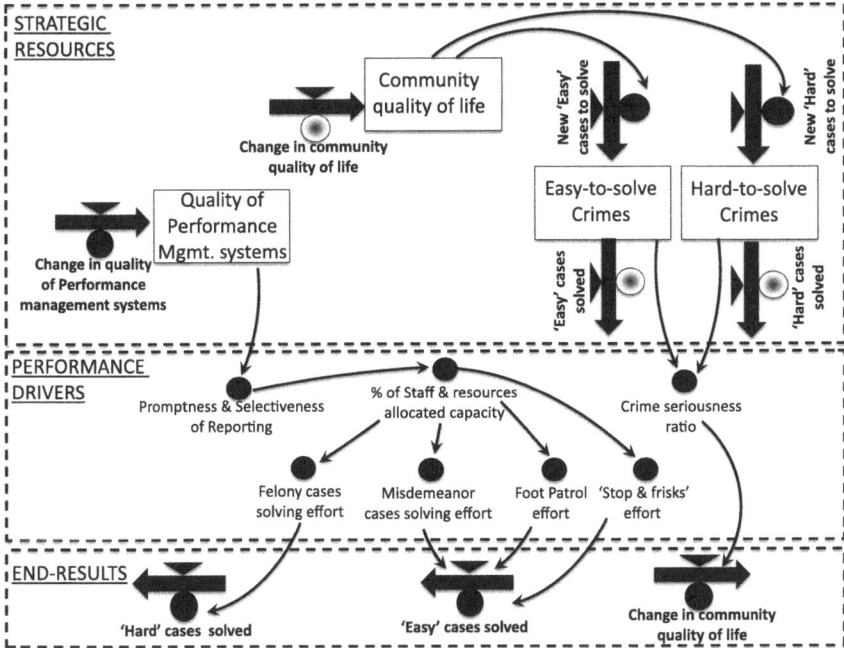

Fig. 4.17   Improving policies through DPM

capacity allocation in solving hard and easy crimes. Balancing the mix of achieved end-results may contribute to the more sustainable development of a public sector organization and the local area. This is shown in Fig. 4.13 through the outcome measure for "change in community quality of life." This variable is, in turn, affected by a performance driver, the "crime seriousness ratio," which is significantly affected by the stock of "hard-to-solve" crimes as a share of total crimes.

4.6   Applying Dynamic Performance Management to Performing Arts: The Case of Municipal Opera Houses

Nonprofit art firms, especially those in the performing arts, have to deal with a difficult problem: their costs tend to rise over time more rapidly than revenues, thus generating a widening budget gap. This is why a key problem in performing arts is on how demand can be increased in order to raise revenue, and what performing arts firms can do to lower their production costs (Heilbrun and Gray 2001).

The argument can be summarized as follows: it takes 70 musicians as much playing time to perform a Beethoven symphony today as it did when the piece was written (Bianchi and Cognata 2010). However, at the same time, new productivity-enhancing technologies have created huge increases in labor productivity in other sectors of the economy. Now, since wages tend to rise with productivity all over the economy, the upward pressure in the 'progressive sectors' determines rising wages also in the 'stagnant sectors' (the orchestra). But since musicians' productivity has not changed, higher revenues do not compensate these higher wages in orchestra. The result is that, over time, costs rise relative to revenue.

According to this explanation, the only way for a nonprofit performing art firm to deal with a widening budget gap is to increase external resources, for example by actively seeking private grants and donations and/or asking for more governmental funds.

Besides the "cost disease", other arguments have been invoked to justify public support to artistic activities. The oldest and most often invoked argument is that art is a public good (Frey and Pommerehne 1989). According to this perspective, performing arts do not only generate benefits for those who pay an explicit price for them, but also for all other citizens, who do not necessarily wish to contribute voluntarily to art production and culture preservation. For instance, museums, opera houses, orchestras and art festivals attract tourists and visitors and generate spillover effects on hotels and restaurants.

DPM can contribute to support performing art firms to deal with the described "cost desease", and to combine efficiency with effectiveness.

An example of a "dynamic" BSC developed for a large Italian public theatre (Bianchi and Cognata 2010) will be shown in this section.

The "Massimo Theatre" is the third largest public opera house in Italy. It employs more than 400 people and has a budget of about Euro 42 million. More than 90 % of the revenue comes from public subsidies; the remaining 10 % comes from the box office and the sale of other services. Salaries for full-time employees represent more than 70 % of production costs.

In the fiscal year 2004 the Massimo Theatre was in a very difficult financial state: losses of Euro 1.5 million, short-term bank debts of about Euro 26 million and production at its minimum. A new management took over at the beginning of 2005 with the aim of preparing and implementing a 3-year plan to restore viable financial conditions.

As previously remarked, increasing the share of earning income relative to total operating costs is essentially the main option for the opera management to deal with its structural earnings gap. This goal can be pursued by either increasing revenues or by reducing operating costs. On the latter side, the new management of the Massimo Theatre negotiated a new contract with the labor unions on a flexible basis, and negotiated a bank mortgage to reduce financial costs. On the former side, the strategy followed was to maximize the number of performances, given the structural and organizational constraints.

The first constraint is that the theatre does not have enough facilities for orchestra and stage rehearsals. Since one needs to use the theatre also for the rehearsals, this limits the number of days one can uses it for performances. This is especially true when production is organized according to the Italian system of *stagione* (the season): a certain number of performances over several weeks, followed by a rehearsal period for the next production after which that opera is performed for the next several weeks, and so on. The typical "season" at the Massimo Theatre is usually 10–11 productions, 8–9 of which are operas and 2 ballets. Opera productions are divided in three groups: at least 3 blockbuster titles (each of which is staged for 8–12 performances), 3–4 repertory titles (7–8 performances), and 1–2 rare and/or contemporary operas (4–6 performances).

Since clients subscribe for the whole season, a mix of repertoire is charged for in the package and this can 'cross-subsidize' less popular repertoire with popular operas. Each season, of the whole number of staged operas, usually the house produces 3 new operas; the remaining are rented from other houses and/or re-used from the Massimo Theatre's stock of previous productions.

In order to support planning and performance improvement at the Massimo Theatre, a program implying the development of a Dynamic Balanced Scorecard was started.

A simulator was built to frame the logical sequence of the planning process for opera production, as previously described. A three-dimensional analysis has been done, i.e.:

1. The *opera mix by type* (blockbuster vs. repertory vs. contemporary/rare) is planned, not only in percentage but also in absolute terms.
2. For each type of opera, *titles by source* are planned. For instance, it is planned how many of the blockbuster operas will be made up by 'new', 'rented', and 're-used' titles; and the same is for repertory and contemporary/rare operas.
3. The *number of performances* by opera type (blockbuster vs. repertory vs. contemporary/rare) is then planned.

The planning time horizon is 3 years. Therefore, each year the number of titles and performances to stage for the coming and the next 2 years is planned. Planning in advance the next season is important to properly manage rehearsals and the acquisition of resources; among them, human resources are the most crucial. In fact, particularly concerning blockbuster performances, it is critical to promptly contract best performers (so called "superstars"), in order to pursue a well-balanced quality/cost ratio in the delivered service.

The SD model also supports an assessment of the financial sustainability of the planned seasons.

Though the planning time horizon is 3 years, the simulation time is 6 years. This is due to the need to allow the user to figure out over a longer time span the possible effects (i.e. the key variables' behavior) of adopted decisions in the coming years. This also allows decision makers to simulate—according a 'rolling plan' mode—how past decisions might require future adjustments, based on the outcomes they are currently generating.

Therefore, by using the simulator, managers are able to adjust and better calibrate their strategies concerning the above three viewpoints, and so they are enabled
to review and discuss them. This contributes to their learning processes and to the
quality of communication.

Figure 4.18 shows three main stock-and-flow diagrams embodied in the model,
concerning the three aforementioned viewpoints that are relevant for opera planning
(i.e. opera mix by type; titles by source; performances by opera type).

The first two structures are depicted as aging chains: the final flow "Blockbuster
titles staged in the current year" depends on the ability of the firm to implement its
3 years plan (Fig. 4.18a). Lack of resources would have a final impact on such

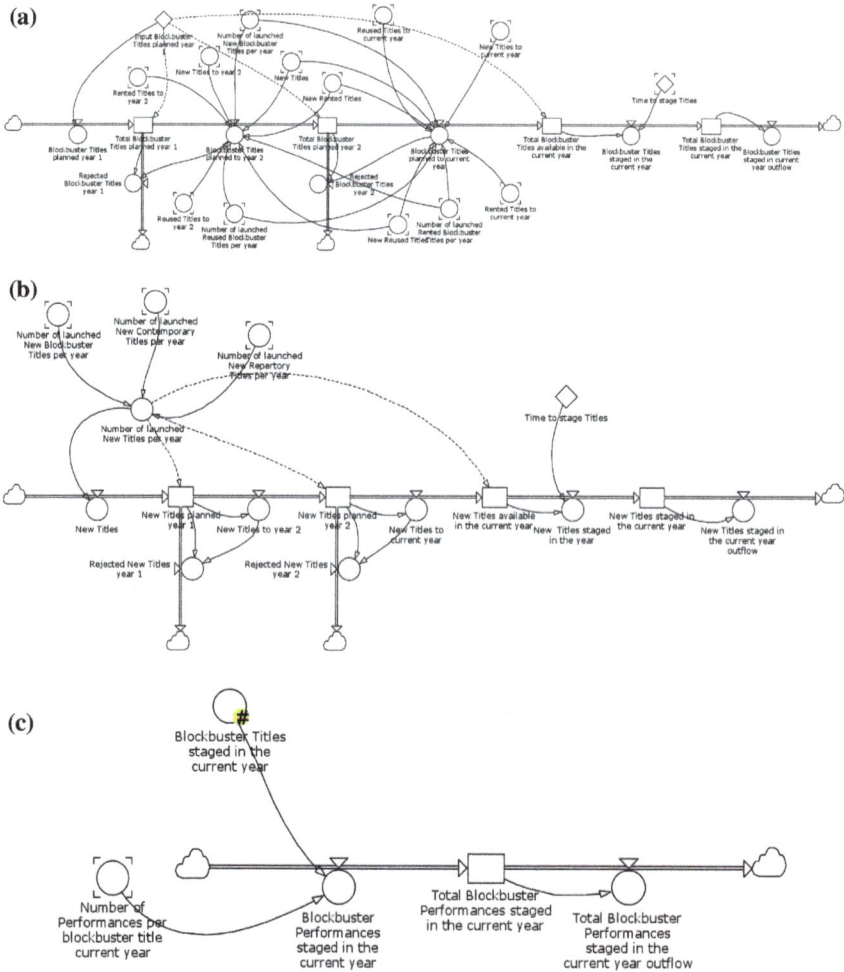

Fig. 4.18 **a** Planning opera mix by type. **b** Planning titles by source. **c** Planning performances by
type

flow, both in volume and in quality of performances. Such quality is also affected by the opera mix by type. On this regard, Fig. 4.18c shows how the same flow determines the number of performances by type.

The three sequential stocks (and corresponding outflows) relate to the planned season for the years (from left to right) n + 2, n + 1, n, where the last year is the current (i.e. the coming year).

The stock-and-flow structure depicted in Fig. 4.18a determines the number of titles by source (i.e. new, rented, reused), depicted in Fig. 4.18b. Here, the flow "New titles to year 2" in turn affects (together with the planned 'rented' and 're-used' titles) the flow of "blockbuster titles planned to year 2".

Therefore, the three subsystems are strictly related to each other, and depend on the decisions made by managers through the support of the simulator control panel. As it is possible to see from Fig. 4.19, such panel depicts the same aging chain structure previously described. Therefore, the planner is asked to define the number of titles per type and related mix (new, rented and reused). The user is also asked to set the number of performances per each type of title in the next 2 years, and (if applicable) the change in the number of performances per title planned for the current year.

Other subsystems of the simulator portray the dynamics of the company's human resources, the customer base, and financial resources. As said, such resources feedback into the opera production and performance subsystems, since they may either support or hinder the implementation of plans for opera staging.

The DBSC chart embodied in the model portrays main indicators related to the four BSC perspectives. A section, concerning the 'customer' and 'internal processes' perspectives, is depicted in Fig. 4.20a, b. Both figures embody main

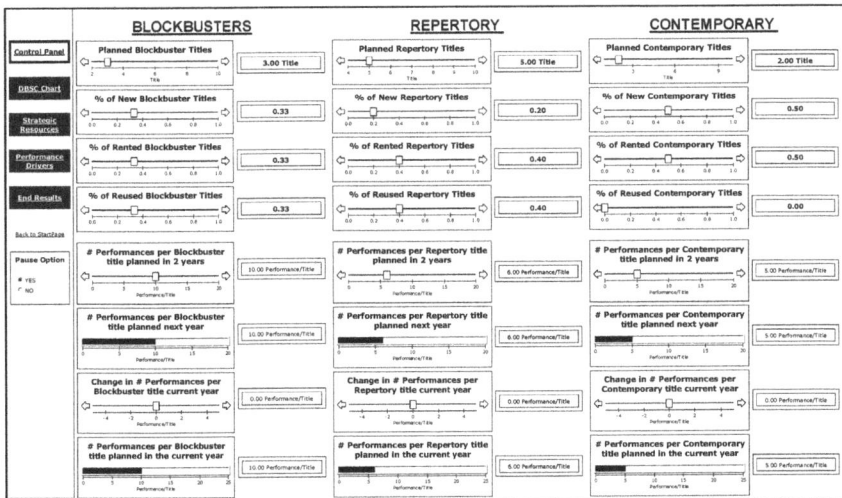

Fig. 4.19 The DBSC control panel

(a)

OBJECTIVES	MEASURES	TARGETS	CURRENT SITUATION
Balancing diversity with repertory	% Repertory titles staged in the current year on total	0.50	0.50
	% New titles on total in the current year on total	0.30	0.30
	% Contemp. & rare titles on total in the current year on total	0.20	0.20
Enhancing the volume of offered operas (titles)	Volume of titles staged in the current year	10.00 Title	10.00 Title
	Volume of titles available in the current year	10.00 Title	10.00 Title
	Volume of titles to stage next year	10.00 Title	10.00 Title
	Volume of titles to stage in 2 years	10.00 Title	10.00 Title
Consolidate our customer base	% Change in subscriptions	0.10 per yr	-0.20 per yr
	% Change in Young Customers	0.08 per yr	0.00 per yr
	% Change in Mature Customers	0.03 per yr	-6.67e-3 per yr
	% Change in Elderly Customers	0.05 per yr	0.10 per yr

(b)

OBJECTIVES	MEASURES	TARGETS	CURRENT SITUATION
Designing productions well ahead of scheduling, to ensure contracting best performers	Time to launch New titles	2.00 yr	2.00 yr
	Time to launch Rented titles	2.00 yr	2.00 yr
	Time to launch Reused titles	2.00 yr	2.00 yr
	% best performers to contract	0.17 per yr	0.17 per yr
	Number of contracted 'Superstar' performers in the current season	96.30 characters	96.30 characters
Structuring offer in a way that a large part of the season is triggered by a few 'blockbusters'	% Blockbuster titles in the current year	0.30	0.30
	% Blockbuster performances in the current year	0.43	0.00
	% Blockbuster titles planned next year	0.30	0.30
	% Blockbuster performances planned next year	0.43	0.43
	% Blockbuster titles planned in 2 years	0.30	0.30
	% Blockbuster performances planned in 2 years	0.43	0.43

Fig. 4.20 a Performance measures related to the 'customers' perspective. **b** Performance measures related to the 'internal processes perspective

objectives, performance measures, related targets as well as the current level achieved for each of these targets. While performance targets are input variables, and therefore the corresponding value can be changed by the planner at the beginning of the simulation, the 'current situation' values are results of the model, and for this reason have been emphasized in a box with a different color from targets, since they cannot be changed. Comparing targets with current results at each simulation stage and for each set of strategic objectives supports decision makers to make judgments on the way the system responds to the adopted policies.

In another section, the simulator separately portrays, through time graphs, the behavior of the strategic resources affecting performance indicators. It also shows in separate sections how such measures can be distinguished into performance drivers and end-results. Also these two other sections consist of time graphs.

Figure 4.21 offers a synthetic representation of how planning opera production may significantly affect the future growth of the theatre.

In particular, the positive loop 'R1' shows how an increase in available resources may allow the management to plan for more titles and performances, leading (other conditions being equal) to higher revenues, liquidity and other available resources to foster future growth. If the loop 'R1' is the growth engine of the theatre, the negative loop 'B1' is a potential limit to such growth. In fact, the more opera titles (and performances) are planned, produced, and staged, the more resources will be consumed.

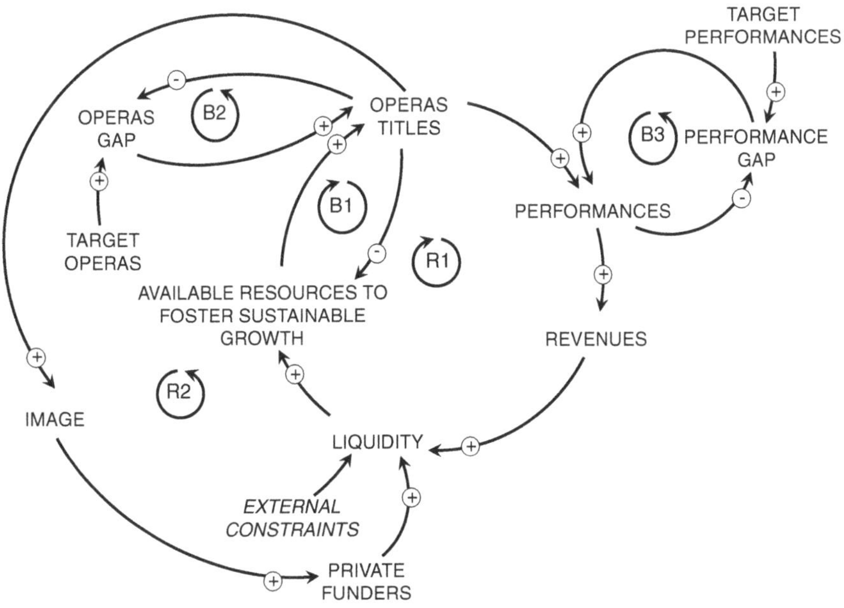

Fig. 4.21 Effects of opera production on financial results and company image

If one particularly considers that opera production planning is characterized by a 3-year time horizon, this means that such phenomenon will reduce (other conditions being equal) the resource endowment to fund the next year plan.

The loops 'B2' and 'B3' refer to the achievement of targets related to planned operas and performances. In fact, the higher the gap from the current to the target number of operas, the higher the number the opera titles to plan will be. This will, in turn, decrease the gap in opera titles. The same is for the planning of performances. The 'R2' loop shows the effect of opera planning on image and liquidity. A higher number of planned and staged operas (and performances) will increase image. This will help the theatre to attract financial resources from private funders. This will increase liquidity and the available resources to foster sustainable growth.

The loop 'R1' portrayed in Fig. 4.22 is similar to the same previously commented about Fig. 4.21. In this case, the effect of perceived quality as a main driver of customer loyalty is remarked. In fact, a higher production of opera titles will lead (other conditions being equal) to a higher perceived quality by customers. This will turn into a higher image, customer base, and liquidity. More financial resources will allow the company to pursue further growth. This will be possible by both increasing the number of opera titles/performances and increasing the number of contracted superstars that will perform blockbuster and repertory operas. This last option will also allow the firm to strengthen growth (reinforcing loop 'R2').

There are a number of balancing loops to take into account, also in this regard. In fact, an increasing number of operas planned in a season will raise costs. This—

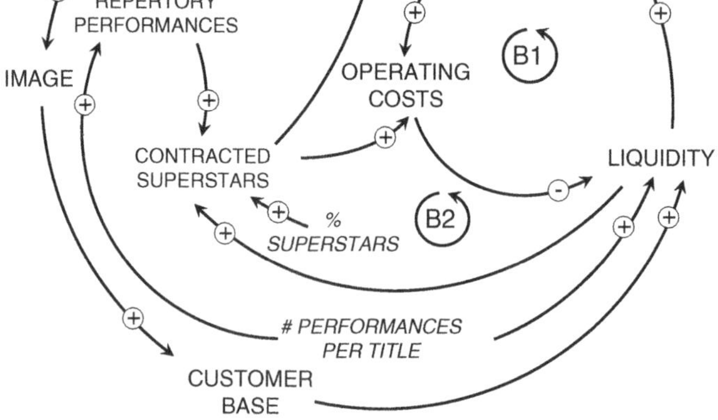

Fig. 4.22 Effects of opera production on perceived quality, customers and financial results

other conditions being equal—will reduce liquidity to plan for the next seasons (loop 'B1'). Likewise, an increase in contracted superstars will also raise operating costs and determine a second potential limit to growth (loop 'B2').

The DPM chart in Fig. 4.23 provides a synthetic view of the effects on performance generated by policies aimed to affect: (a) the time to launch titles; (b) the percentage of staged titles on the total planned, and (c) the personnel costs as a percentage of the contribution margin.

As previously remarked, planning in advance the next season is important to manage rehearsals and the acquisition of resources. In particular, it is critical to be able to offer enough blockbuster titles and to timely contract "superstar" performers. As Fig. 4.23 shows, the time to launch titles affects quality through: (1) the number of blockbuster titles on total, and (2) the quality of contracted performers.

So, the "time to launch titles ratio" is a fundamental performance driver affecting another performance driver, i.e. the "performance quality ratio".

In order to reduce the time to launch titles, management must improve and properly combine two main strategic resources, i.e.: personnel and liquidity.

A second important performance driver illustrated by Fig. 4.23 is the percentage of staged titles on the total planned titles. As previously remarked, this measure affects the number of performances staged by the opera house (which also affects image). An increase in the number of staged performances in a season (end-result)

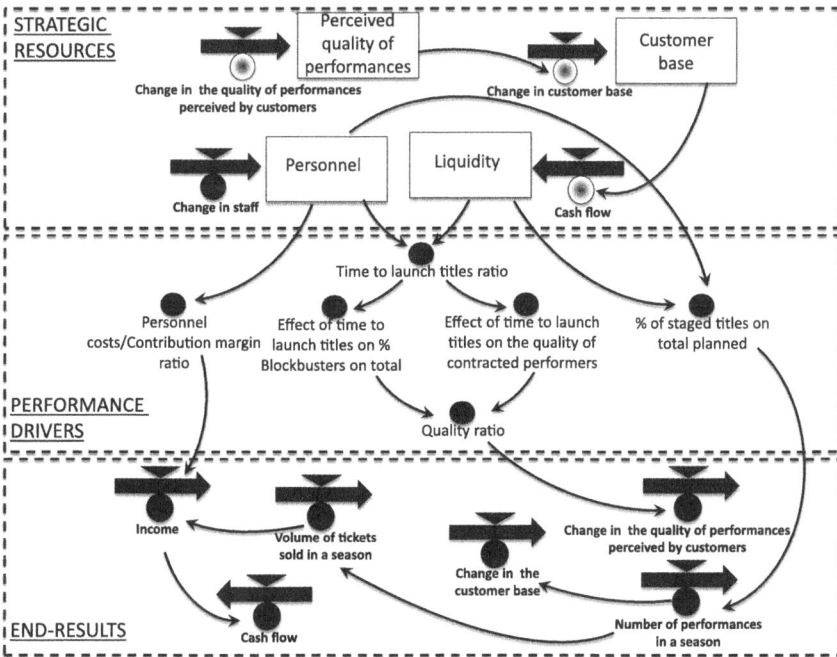

Fig. 4.23 Effects on performance generated by policies aimed to affect: **a** the time to launch titles; **b** the % of staged titled on total planned, and **c** the personnel costs % on contribution margin

affects two other end-results, i.e.: the change in the customer base, and the number of tickets sold in a season. The latter measure, in turn, affects income and cash flow.

Figure 4.23 also shows that the quality ratio affects the change in the perceived quality of staged performances (end-result), that changes the stock of perceived quality (i.e. a fundamental component of customer satisfaction, corresponding to a strategic resource generated by company internal routines). The word-of-mouth effect generated by such strategic resource may corroborate the acquisition of new customers generated by the number of performances offered in a season.

The third performance driver illustrated in Fig. 4.23 (i.e. the Personnel costs percentage on contribution margin) underlies a possible trade-off. In fact, increasing the number of titles and reducing the time to launch them requires more and/or skilled staff. This may provide a basis for generating growing revenues (income and cash flows), and competitive performance. However, increasing staff may imply a rise in fixed costs, which would erode the contribution margin, and therefore income and cash flows. Therefore, as previously remarked, efficiency in managing personnel costs is critical to pursue financial and sustainable competitive performance.

4.7 Adopting an Inter-institutional Perspective and an "Objective" View of Dynamic Performance Management for the Effective Implementation of User Satisfaction Programs

4.7.1 Introduction

Quality in delivering services to citizens and the community has been recognized as a major issue challenging performance in the public sector. However, the design and implementation of sustainable user (or customer) satisfaction programs is still a controversial theme.

Conventional approaches to this topic mainly aim to gather data from public sector user perceptions, which are captured through different methods (i.e. interviews, focus groups), and then treated through statistical techniques. Though such data can be very useful in knowing possible improvement areas and achieved results in public service delivery (as perceived by the interviewees), they represent only a partial input to decision makers in order to design sustainable user satisfaction programs. In fact, they often do not primarily take into account: (1) *how* a given set of provided services is delivered; (2) *who* is accountable for the achievement of results directly and indirectly associated with service provision, and (3) *what* is the system ruling the distribution of roles, duties and competences affecting a given public service.

Such programs usually focus only on the front-line activities performed by those institutions operating in the *last mile* of a much longer service delivery system, in the public sector 'value chain'. This runs the risk of adopting a too-bounded viewpoint, which may generate a feel of disappointment, and lack of motivation in front-office units. Such feelings are often associated, for instance, with: lack of coordination with upward public sector institutions; inconsistencies in the 'roles/responsibilities' system, or in process design and management, or in results orientation—particularly in the back office units.

Supposing that performance improvement regarding user expectations could result only from efforts produced by those institutions operating in the bottom and front-line level can be misleading. In fact, such institutions are, in turn, 'clients' of other public institutions that operate at higher jurisdiction levels.

This section will outline a non-conventional view of 'customer satisfaction' in public sector service provision. An inter-institutional perspective and an "objective" view of DPM are proposed in order to map the overall value chain leading to the final product delivered to users.

4.7.2 Broadening System Boundaries to Deal with User Satisfaction Programs

If one considers a "product-service" as an outcome of a value chain encompassing different institutions (all of which are loosely coupled to each other) the formulation of user satisfaction policies requires that the investigated system's boundaries be broadened.

For instance, in Italy, this is the case of performance improvement policies in the execution of infrastructure funded by the European Union "Regional Operational Program". In this regard, each municipal administration operates as the last segment of a long value chain, which is made up by several units operating in different institutions, with ruling and monitoring roles, most of them are inside a Regional Ministry of Infrastructure.

Likewise, concerning immigration policies, social and family assistance services in the fields of council housing, education, or employment support are mutually linked and affected by European, national and regional policies. One can see a similar pattern in the policies undertaken by a Regional Ministry of Labor, concerning Employment Exchange[3] and training measures, both aimed to deal with youth unemployment rates.

Particularly, unemployed youth education policies disclose interdependencies between the national and the regional Ministry of Labor. Other interdependencies are between the Regional Ministry of Labor and various 'brokers' such as: private

---

[3]In the Italian system, the 'Employment exchange' provides a national on-line data transmission system, aimed to facilitate the free match between demand and supply of labor. It is an open, transparent, free, and updated system, made up by a net of regional nodes, through which citizens, enterprises, public administrations, and authorized 'brokers' can have access to share resources, information, and inquiries on labor supply.

training centers, enterprises asking for qualified manpower, and graduates searching for employment, and receiving training. In such a 'service delivery' system, unemployed graduates are the *final clients*, while private training centers are *internal clients.*

Similar arguments can be related to public policies on: water supply, waste management, land and energy. In the European context, the Act n. 387/03 (art. 12), under the European Directive n. 77/01, delegated to the Regions (second level administrations in a State) the tasks of authorizing applicants (e.g. enterprises) for the construction of renewable energy production plants. This authorization must be in conformance with laws affecting various issues such as environment and landscape, historical and artistic heritage. For instance, the "Industry" department of the Sicilian Region must call a summit Conference of Services for each received application, within 30 days from its receipt. It must also conclude the single authorization process within 180 days.

The range of institutions potentially involved in such conference of services is significantly broad. It covers different regional ministries (such as 'Land and Environment' and 'Public Works'), Superintendence for Cultural Heritage, Forest Inspectorate departments, Municipalities, Industrial Development Areas, Local Health Agencies, and other various institutions, like the Italian Company for Air Navigation Services (ENAV), the Civil Aviation Authority (ENAC), Provinces, the State Property Agency, the Air Force, other State Ministries (e.g. Communication, Defense). In such a complex and multifaceted domain, the "Mines and Energy" Service of the Industry department must operate as a *unified desk*. In this regard, the capability of such Service to detect criticalities and inform and address the 'client' (concerning the identification of needed or missing documents in the application), is a first important factor affecting performance. Another important factor is the prompt and right identification of institutions to call in the summit Conference of Services.

This particularly requires a good knowledge of laws and a capability to analyze and diagnose problems, and to correctly bring them back to legislative frameworks, rules, and related current interpretations. Even the capability to outline a reliable time schedule of the needed conference of services calls is critical. Last, but not least, the capability to deal with 'political-cultural' problems related to change resistance by different involved public sector stakeholders plays a significant role in affecting performance.

4.7.3 Mapping "Products" and "Users/Clients"

In all the examples mentioned above, the identification of "products" and "users/clients" provides an important key to start a program to affect performance from a user satisfaction perspective.

Such a perspective must first identify the 'user-client', to whom the final products emerging from the fulfillment of administrative tasks are addressed (Holzer and Kloby 2005). Starting from such 'clients' and final 'products'—and then moving backward through the value chain (Bouckaert et al. 2005) which links different loosely coupled institutions—allows one to identify different internal clients in the public administration system. Internal clients are those public agencies or organizations that receive from other units in the public sector a given range of 'products' from the fulfillment of their own administrative processes. They deliver, in turn, the products arising from the execution of their own administrative tasks to the benefit of down-stream public agencies, with a view to reaching the citizenry, i.e. the final users.

The conceptual framework here described implies that—if one refers to a given *final* "product"—it is possible to identify, by moving backward, a *system of products* resulting from the fulfillment of administrative processes by each decision unit whose only "clients" are *internal* to the public sector.

For "products" which are both delivered to the benefit of *external* and *internal* "clients", the identification of factors affecting performance and customer satisfaction requires an analysis of: (1) underlying processes and activities; (2) involved responsibility areas; (3) related available policy levers, and allocated resources; (4) performance indicators.

With a goal of gradually moving from synthesis to analysis, it is necessary—for each institution—to first outline the *macro processes* carried out by the respective responsibility areas.

For each group of macro processes underlying the delivery of a given output, at least one *intermediate* "product" must be identified. Such "product" is an autonomous result of administrative tasks fulfilled by 'back office' units to the benefit of an *internal* "client". Performance in delivering an *intermediate* product will affect the performance of the *internal* "client" who will receive it, and will—in turn—influence the performance of other internal clients who are sequentially located along the *value chain* leading to the delivery of the *final* "product".

In order to make this analysis selective, it is worth focusing attention on the top-middle management areas in each institution, i.e. on the *second level* units in a department. They provide a crucial area for performance improvement. Decisions made by managers having the responsibility of such units are a good compromise between the need for synthesis and coordination of the detailed activities accomplished by lower level units, and the need to focus the specific processes behind the acquisition and deployment of strategic resources needed to affect performance. The focus on these organizational levels may foster the empowerment and accountability of managers operating under the supervision of the department's director. A missing analysis of such processes and of their critical issues can be an obstacle to the identification of the "levers" on which managers may act to affect the drivers impacting on the end-results generated by fulfilled administrative tasks. It would also hamper goal setting and negotiation, budgeting, the assessment of results, reward and career systems.

The identification of *macro processes*, of related *intermediate* products to the benefit of *internal* customers, and performance measures, strategic resources and policy levers on which decision makers must focus their attention, provides the first.

A second step for implementing DPM to improve user satisfaction in the public sector requires that—for each *macro process*—different single processes are made explicit. The result of each single process is an *output* of administrative tasks. This output can be referred as a 2nd level *intermediate* product, i.e. a result which is an instrument—often together with other 2nd level *intermediate* products—for the attainment of a 1st level *intermediate* product, as previously defined.

Figure 4.24 depicts *intermediate* products as outputs of administrative tasks fulfilled by *second level* units in a department of a public sector institution.

If we consider the overall *value chain*, one should take into account three sequential areas for improvement:

- The *final* "product". It underlies *outcome indicators*, related to the overall value chain, to the performance of several institutions, involved with different roles in the delivery of the considered service to citizens;
- *1st level* intermediate "products". It underlies performance drivers, which could be named as *second layer* indicators;
- *2nd level* intermediate "products". It underlies performance drivers, which could be named as *third layer* indicators.

Figure 4.25 shows how outlining activities, processes, products and related performance indicators, identifies two opposite flows: an *information flow* and a *physical* flow. The first one allows the planner to map the above mentioned *value chain*, by moving backward from the final "product" to processes, and—from them —to activities. The second one describes how the public administration system generates or destroys value, in delivering a given service. Therefore, it supports the identification and calibration of performance indicators, and related intervention fields, in a user satisfaction perspective.

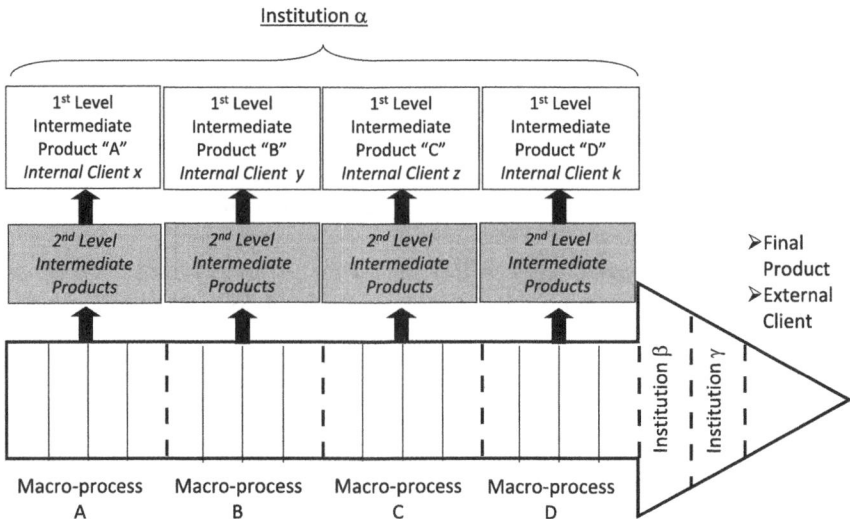

Fig. 4.24 *Final* products and *external* clients, and related *intermediate* products delivered to *internal* clients through the fulfillment of macro-processes inside an institution "α"

Fig. 4.25 Activities, processes, "products" and performance measures related to the value chain for public service delivery

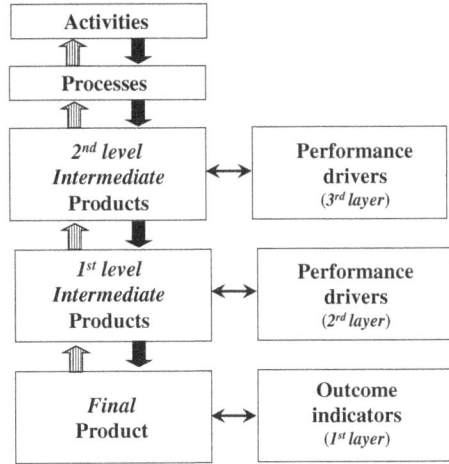

4.7.4 Implementing User Satisfaction Programs Through DPM

Implementing the approach here described requires a selective field analysis, to map the institutions that play a role in the *value chain* underlying the delivery of a given product (or a group of related products).

Since the administrative activity of any public institution is to a significant extent governed by law, this analysis needs to first outline the legislative framework and the formal system of administrative rules that comprise institutional system at hand. The formal rules providing the backbone of such system can be considered either as a constraint or a *lever* to foster change and performance improvement. In the first case, one sees law as an external and unchangeable factor that requires compliance by any customer satisfaction program. In the second case, one takes a longer view, with the possibility to revise the legislative framework in order to redesign the institutional system and the rules that govern it.

It is important that such field analysis be conducted following the value chain backwards. A reverse perspective implies a previous detailed analysis to build a database of activities, roles and tasks fulfilled by each unit operating in an institution, to cover the *universe* of heterogeneous competencies and duties attributed by law and administrative rules. According to such an approach, a massive collection of data is often done through interviews and questionnaires delivered to the employees working in each unit at all the organizational levels, with the intent to know how they spend their working time. This analysis usually results in a collection of data that are barely consistent and symmetric with each other when compared against one another. Therefore, this approach runs a serious risk that it may be unfocused and too vague and useless to implement a performance improvement program. Although it may provide a complete picture of what each single unit in a given institution actually does, it might not be very helpful in detecting delivered 'products' and related 'clients', as well as the processes through which they are delivered, and the proper performance measures to adopt in order to improve performance.

A *top-down* approach rather implies a more selective detection of the needed data to feed information needs. Moving backwards in such analysis, i.e. from the *final* "products" delivered to *external* "clients" at the end of the relevant *value chain*, firstly focuses the analysts' attention over the *macro-processes* through which *intermediate* products are delivered to *internal* "clients", inside or outside a public sector institution. Such analysis can be done through semi-structured interviews firstly addressed to department directors, then (more extensively) to managers leading the *second level* units, under their supervision.

In particular, interviews should gather useful data about: (1) available resources in each responsibility area; (2) the flexibility in their use; (3) possible constraints in the acquisition of further resources (in addition to the available ones); (4) bottlenecks and rigidity factors for performance improvement due to the activity of other units, located backwards in the same or other institutions; (5) different factors (waiting time, number of faults, allocated working time) about which it could be possible to improve the efficiency and effectiveness of management processes and the quality of delivered "products". The above data are also useful to calibrate performance standards in a budgeting setting, and to support periodical analysis of variances between actual and budget data, on which the control system feedback mechanism is based.

When possible, such interviews should also be supported by periodic workshops —involving different decision areas, operating in the adjacent processes or even in macro-process different from each other. They could also involve decision makers from different institutions—from the private sector and representatives of *final* 'clients' too—belonging to the *value chain*. Participation in a workshop by people from different institutions (and, more generally, different decision areas) can stimulate dialogue, a better knowledge of respective goals, constraints and mindsets, the identification of problems and related possible solutions, and the search for continuous improvement.

To this end, it is necessary that such workshops be designed and implemented as proper working sessions whose purpose is knowledge and strategic organizational learning. Therefore, a *learning facilitator* should be the leader of such sessions. He or she would have to use a scientific method to provoke and manage *tension* in the participants group towards the detection of crucial issues underlying the delivery of the "product" or the group of products taken into account. The process of analysis should be focused on investigation of the causal factors affecting performance, moving backward from end-results to performance drivers and strategic resources, and the various available policy levers.

4.7.5 Case-Study: Regional Department of Infrastructure

This section will illustrate the results of a project that was carried out with a group of managers of an Italian regional department of infrastructure, with the goal to improve efficiency in the exploitation of the measure n. 5.02 of the European Union (EU)—Regional Operation Program (ROP), for the so-called "objective 1" regions. The EU "objective 1" aims to promote economic growth and to enhance long-term competitiveness in less developed areas. More specifically, in order to tackle social marginality, the program aims to recover historical urban centers through the construction or the modernization of parks, buildings, roads, schools, etc.

The local section of the Court of Auditors had been stigmatizing the regional public sector for its low capability to efficiently use such funding. During the seven years financed by the EU through the program, the Region had been able to use only about the 50 % of the total funding, while other Italian Regions had been able to exploit around the 70 %.

Though spending does not measure outcomes, such statistics were providing an important symptom of a low aptitude of the regional public sector to support the building and delivery of proper infrastructure to the community.

The project aimed to outline: (1) the main 'products' associated with the fulfillment of administrative tasks by the department, (2) the underlying macro-processes, (3) the involved responsibility areas, and (4) performance indicators.

Four macro-processes and different levels of intermediate administrative products were identified.

The main goal of **macro-process (1)** is to reduce the time it takes by the department to publish the Decree authorizing the issue of the call for tenders related to the public works projects by the Municipalities that will benefit from the EU funding.

Such decree is a "first level" intermediate product delivered by macro-process (1). Related to it, two "second level" intermediate products were identified, i.e.: (a) the call for tenders that the regional administration issues to invite Municipalities to submit projects to be funded through the EU funds, and (b) the regional decree approving the list of funded projects.

The main performance indicator related to this macro-process is the "actual/standard" time ratio to issue the decree.

Main activities and responsibilities, related to the fulfillment of this macro-process are:

(a) Call for tenders is issued: Municipalities are invited by the regional administration to submit projects;
(b) Project proposals are received and selected;
(c) A decree approving the list of projects admitted to funding is issued;
(d) Copy of the decree is forwarded to the "Service n. 2" of the regional department of infrastructure (this is a second level unit in the department);

(e) "Service n. 2" may ask each beneficiary municipality to send supplementary documents. In the meanwhile, each funded Municipal administration appoints a project manager;

(f) The department's director signs the decree authorizing funded Municipalities to issue their call for tenders.

The main goal of **macro-process (2)** is to minimize the number of appeals filed by construction firms against the call for tenders issued by the Municipal administrations.

The "first level" intermediate product delivered by macro-process (2) is the drawing up and signature of the contract between the Municipal administration and the construction firm that will implement the infrastructure. Related to it, other "second level" intermediate products were identified, i.e.: (a) the call for tenders that the Municipal administration issues to invite construction firms to submit projects for the implementation of the funded infrastructure, (b) documents forwarded by the "Tenders Office" of the Regional Department to the project manager for the funded infrastructure at each Municipal administration.

The main performance indicator related to this macro-process is the "actual/standard number of appeals" ratio. Also, since an important means to minimize the number of appeals filed against the Municipal call for tenders is considered the standardization level of administrative processes and the quality and size of databases on tenders, appeals and court verdicts, another performance indicator is the ratio between the actual and the standard average time it takes by the regional department to update databases.

Main activities and responsibilities, related to the fulfillment of this macro-process are:

(a) The Municipal project manager for the funded infrastructure sketches the call for tenders based on which construction firms will submit their projects to implement the infrastructure. If the value of the project funding is higher than Euro 1,250,000 the "Tenders Office" of the Regional department will monitor the process of sketching the call for tenders, and will support the project manager for each funded infrastructure at Municipal level.

(b) After the Municipal administration has issued the call for tenders, the "Tenders Office" of the Regional department will receive the documentation from the project manager of each Municipality. Such documents will be submitted to the Project Evaluating Committee, whose members are selected by the "Tenders Office". This office also provides administrative support to the Committee.

(c) After the committee has selected the construction firm that will implement the infrastructure, the "Tenders Office" will draw up the minutes of the selection process and will forward them to the project manager at each funded Municipality.

(d) The Municipality and the selected construction firm will sign the contract.

The main goal of **macro-process (3)** is to minimize the time it takes to implement the infrastructure. This target is considered as an outcome that the regional department of infrastructure is not able to directly affect, since project management is implemented at municipal level. So, the regional department mainly focuses on targets impacting on the time it takes for the municipal administrations to spend the EU funds allocated to build infrastructures.

The "first level" intermediate product delivered by macro-process (3) is the funding decree issued by the regional department of Infrastructure to the Municipalities that will benefit from project funding.

Related to it, the technical certification of expenditure was identified as a "second level" intermediate product.

The main performance indicator related to this macro-process is the "actual/standard" time to issue the credit order, through which funds are made available to the Municipal administration to implement the infrastructure. Other performance indicators through which the regional department may monitor the efforts made to prevent and correct problems that may slow down the process leading to the issue of the funding decree are related to: (1) the ratio between task force meetings done/planned to solve problems, and (2) the ratio between supervisory visits to Municipal administrations implemented/planned by the regional department.

Main activities and responsibilities, related to the fulfillment of this macro-process are:

(a) The Service n. 2 of the regional department prepares all the necessary documents to issue the funding decree;
(b) The Municipal project manager forwards to the Service n. 2 a "Technical certification of Expenditure", which summarizes the budgeted costs that will be funded by the first payment to the construction company;
(c) Such documents are checked by the Service n. 2, which support project managers to find solution to possible problems.
(d) The regional department of infrastructure issues the funding decree. Payment for a first fraction of the project is done.

The main goal of **macro-process (4)** is to minimize the time it takes to forward to third monitoring institutions documents related to infrastructure implementation.

The "first level" intermediate product delivered by macro-process (4) are monitoring and reporting documents.

Related to them, statements of account, monitoring summaries, and certificates of payment were identified as a "second level" intermediate products.

The main performance indicator related to this macro-process is the "actual/standard" time to submit reports to third monitoring institutions.

Main activities and responsibilities, related to the fulfillment of this macro-process are:

(a) Municipal project managers periodically forward to the Area n. 2 of the regional department of infrastructure the documents proving payments for

funded expenses (i.e. statements of account, monitoring summaries, and certificates of payment).
(b) The Area n. 2 of the regional department of infrastructure controls and vali-
dates the documents. Also, every two months it submits to the "Planning unit"
of the Central Regional Cabinet a report on the expenses for each project.
(c) It also forwards the same documents to the Service n. 2, to enable it to
undertake the necessary steps to speed up the project execution (e.g. through
meetings with municipal project managers, aimed at detecting and solving
problems).

4.8 Applying Dynamic Performance Management to Local Areas: The Case of Environmental and e-Government Policies

Though local government is a traditional field of research and practice in performance management, the governance of local areas is a relatively new topic. It
primarily requires a focus on a multi-organizational (i.e. inter-institutional) sphere,
where performance measurement/management is expected to foster coordination
among different players to pursue a learning-oriented strategic planning.

Local strategic planning is crucial for sustainable development. It implies the
implementation of 'metropolitan governance', i.e. "the process by which citizens
collectively solve their problems and meet society's needs, using 'government' as
the instrument" (OECD 2000, p. 1).

The concept of competitiveness, when applied to local government, goes beyond
a mere aggregation of the competitiveness of companies or single organization
located in a region. Local areas can be considered as independent agents, which
may compete on a global scale, to attract and retain strategic resources (e.g.: mobile
investment, public funds, infrastructures, companies, population, human capital,
tourism, arts, and global events) that will allow them to further improve their
competitiveness, to preserve or increase quality of life and social wellness (Begg
1999; Jessop and Sum 2000; Lever 1999; Porter 1995). For instance, human capital,
companies, infrastructures, knowledge networks and the transparency of public
sector authorization processes, may further attract new companies, projects, and
skills. The capability of a region to attract, retain and deploy such resources may
foster the acquisition of further strategic resources that cannot be gained through
'market-like' competition such as: quality of life, social capital, citizen satisfaction,
trust in government, and reputation.

However, the governance of local areas is not that simple due to a number of
factors, such as: fragmentation of jurisdictions and lack of coordination among
them, blurred decision-making, chronic difficulties in financial management and
fiscal policy making, lack of accountability and of outcome-driven vision (OECD
2000). Changing attitudes and developing a culture of governance is a pre-requisite

to improve the capability of metropolitan areas to pursue local development (OECD 2001, p. 13). More inclusive and participatory governance approaches should replace traditional "top-down" rule-driven systems (OECD 2000). This implies a shift from "government" to "governance" (Cavenago and Trivellato 2010). In this context, governmental institutions are expected to take an active role as leaders of a change and learning process, implying a constant interaction, not only with other public sector organizations, but also with the civil society.

A number of factors shape the specific complexity of this field.

First, the planning effort is primarily focused on a geographic area. In this perspective, as remarked in Chap. 2, the sustainable development of single organizations is tightly related to the sustainable development of the local area where they are located.

Second, coordination between different institutions and an outcome view to investigate how adopted policies will impact on the performance of the area are needed.

Third, a trade-off analysis (in both and time and space) is needed: a strategy might improve local performance in the short term, but it might also lead to unintended outcomes in the long term. Also, it might improve performance in a given industry to the detriment of another.

Fourth, a feedback perspective is needed. In fact, an input-output view may lead to a bounded identification of the relevant system generating local performance, and therefore to a poor assessment of policy outcomes. This also requires a learning-oriented approach: a cross fertilization between different professional profiles participating into the planning process should contribute to capture the systemic, complex and dynamic structure of the problem context.

A lack of perception of such complexity has been a major cause of the ritual implementation of local areas' governance (Razumeyko 2011, pp. 406–408).

Though urban studies are a tradition in the SD literature (Forrester 1969), the last two sections of this chapter will outline a relatively new stream of research and practice to propose DPM as an approach applied to local governance. This approach builds on the work developed by Ghaffarzadegan et al. (2011) and by Kim et al. (2013), where small SD models were used to enhance public policy and decision-making.

Two exemplary case-studies will be illustrated through a DPM approach. The first refers to environmental and e-government policies. The second refers to a strategic repositioning of a community's ceramic industry. This section will focus on the first case.

The case study of Hammarby, referred to a district of Stockholm, Sweden (CABE 2013), will be described. Hammarby is an admirable example of urban transformation. Formerly a run-down part of town in Stockholm's industrial area affected by heavy pollution problems, it has become an environmental role model in less than a decade. The city council developed an eco-cycle model to integrate environmental results with strategic planning. An ambitious program to recycle waste and wastewater to turn into renewable energy was introduced.

Solid waste is disposed into a vacuum-based underground collection system, which separates it into organic, recyclable and other forms. Organic waste and bio-solids are converted into bio-fuels and combustible garbage is burned to supply electricity and hot water with the use of heat and power plants and thermal power stations. Also a high percentage of nitrogen and phosphorous is retained to be used as fertilizer for local agriculture. The use of recycled water and waste not only has affected a reduction in pollution; it has also transformed them into strategic resources for the production of environmentally friendly electricity.

The initial success of the Hammarby model to bring forward the priorities of integration of urban sustainability, information and communication technology, and energy efficiency led to the recent establishment in 2011 of a Ministry for Information Technology and Energy within the Swedish Ministry of Enterprise, Energy and Communication. This is a notable example of how local governance and e-government can support the coordination of sustainable development policies across different urban organizations and stakeholders.

A DPM model framing such policies is displayed in Fig. 4.26 (Bianchi and Navarra 2013).

The figure shows how waste depletion and the move towards environmentally—produced energy are specific end-results for the project. They allow decision makers to affect waste accumulation and the stock of environmental friendly

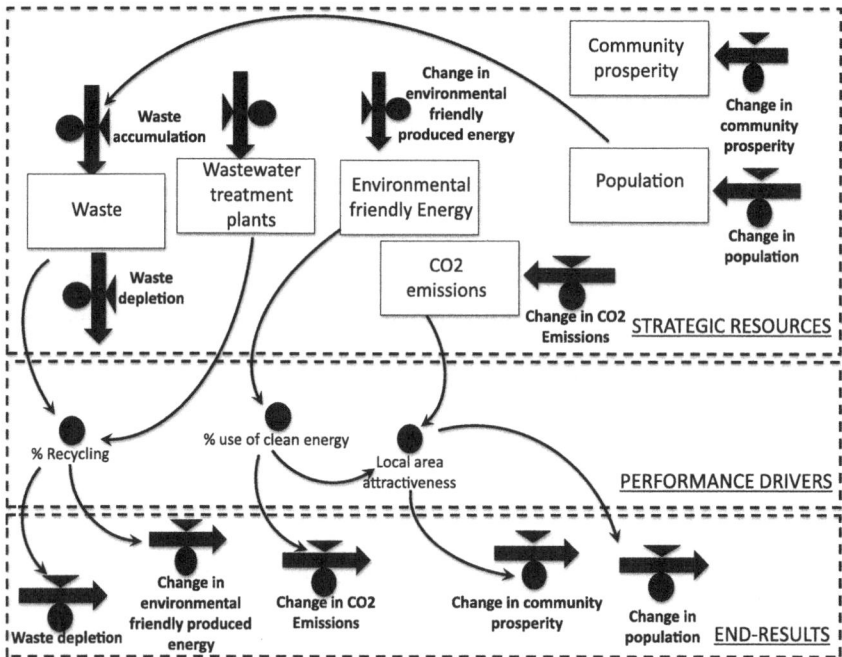

Fig. 4.26 DPM chart portraying the effects of Hammarby's environmental policies on the local area's performance

produced energy. Such end-results are affected by the recycling percentage (i.e. the ratio between recycled and total waste). This driver is, in turn, affected by the investments in wastewater treatment plants (strategic resource) and by the stock of waste.

Figure 4.26 also shows how the stock of environmental friendly produced energy affects a second driver, i.e., the percentage use of 'clean' energy. This driver, on the one side affects the change in CO_2 emissions in the region (i.e. a third end-result); on the other side, it directly affects the local area's attractiveness (i.e. a third performance driver). In fact, a higher percentage of clean energy produced in the area will not only affect the environmental pollution levels, but will also directly generate less expensive electric power available in the region.

The previously mentioned performance driver—"local area's attractiveness"—is also affected by the stock of CO_2 emissions in the urban area.

The outcome of the described policies is related to the effect of local area's attractiveness—other things being equal—on community satisfaction (i.e. prosperity and wealth) and on the stock of population.

This is likely to generate counterintuitive effects in the urban area's performance. In fact, on the one side higher achieved community prosperity will sustain more investments to foster the described policy. This might also imply a stronger and wider level of participation by several decision makers from different public/private sector institutions. This would indicate a higher level of trust and cohesion in the area. Therefore, such first effect of the policy would underlie a growth-oriented reinforcing loop.

However, on a longer time horizon, higher community prosperity might attract more people to live or work into the city. A higher population—other things being equal—might increase the level of waste. This might counterbalance the positive effects generated by the reinforcing loop previously described. Therefore, an increase in population might generate a shift in loop dominance, i.e. from a reinforcing to a balancing loop, acting as a limit to growth.

Policy makers might decide to counteract this loop by exploring alternative scenarios and policies. For instance, they might increase the volume and quality of investments in recycling equipment, with a view to keep stable the percentage of recycled waste. They might also consider levying taxes on accumulated waste. This policy might discourage further increases in population. However, it might level off —or even reduce—the local area's attractiveness, and community prosperity.

The role of e-government policies and systems in the production of the results previously described is not the least important.

The Swedish administrative model, with its independently managed central-government agencies, is considered to have had a major role on the rapid development of digital applications and e-services within the public administration. According to Lind et al. (2009) the historical reconstruction of Sweden's efforts in e-government point strongly towards the requirement for common standards to establish a fully integrated networked public administration. Such standardization has supported the creation of the Hammarby's eco-cycle model.

In 2001, the Swedish government appointed a forum for Information and Communication Technology (ICT) and the environment, under the then Ministry for the Environment. The idea was to create a natural platform for ICTs and ecologically sustainable development. The project was implemented in a working group consisting of representatives from industry, research, the Swedish Environmental Protection Agency, ministries and environmental organizations. One of the group's tasks was to study the potential of ICT use in the development of emerging infrastructures, products and services that are resource efficient and less environmentally harmful. The work continued in the form of the Government's Strategy Group on Information Technology (IT) Policy under the then Ministry for Industry (Bianchi and Navarra 2013). Members of the group and its secretariat staff were recruited from both the private and the public sector.

The monitoring and direct management of energy consumption, an emphasis on services (such as e-commerce, teleworking and e-government) facilitated the emergence of ICT solutions which are energy efficient, such as "thin clients", grid computing and virtualization technologies.

In this context, partnerships between sectors have been introduced to accelerate the development and wide-scale rollout of ICT-based solutions for monitoring, managing and measuring energy-use and carbon emissions in energy-using activities. Finally, smart meters were introduced to enable feedback to be given to the end-user, and instructive online solutions that encourage energy efficiency, eco-visualization and energy simulations, to promote a more responsible behavior in energy use and information transparency.

Last, but not least, the Swedish government promoted the development of knowledge on energy issues in the construction and property sector in a wider sense. A cooperation project was implemented by the national government with other public authorities and the construction industry, involving training programs focused on business operators in the industry.

Figure 4.27 presents a DPM model of the described e-government policy in Sweden, with a specific focus on environmental issues.

The figure shows how e-government policies improved the transparency of information (strategic resource) that was made available to different stakeholders (e.g. enterprises, non-profit organizations). This improved two main performance drivers, impacting on the private sector participation to public governance projects, i.e.: information capillarity (i.e. its availability to a wide range of stakeholders) and information speed (i.e. prompt availability of available information).

Enhanced private sector participation further contributed to improve information transparency (a reinforcing—growth oriented—loop). It also improved the local areas' attractiveness (performance driver), since more private sector organizations were attracted to start projects in those areas (end-result). An improvement in regional attractiveness was also facilitated by a reduction of "grey areas" in public decision-making, i.e. a lower uncertainty for private investors in interacting with the government.

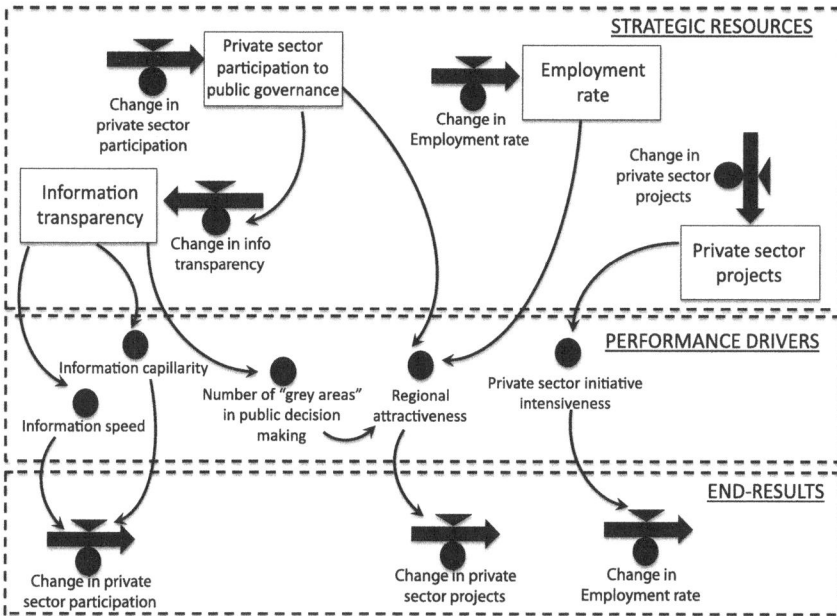

Fig. 4.27 DPM chart portraying the effects of e-government policies in Sweden

The figure also shows that an improvement in the stock of existing private sector projects in a region (strategic resource) boosted the intensiveness of private sector initiatives, which increased the employment rate. An improvement in the average employment rate (strategic resource) in a region further boosted its attractiveness.

4.9 Applying Dynamic Performance Management to Local Areas (Continued): The Case of Ceramic Industry Revitalization Policies

In this section a case-study of ceramic industry revitalization policy making will be discussed, with the goal to describe the weaknesses of a traditional approach to local strategic planning in framing sustainable strategies and supporting policy makers to implement them. Based on such analysis, the benefits of DPM applied to local strategic planning will be illustrated (Bianchi and Tomaselli 2015).

The case-study refers to the municipality of Caltagirone, a small town located in Sicily (Italy). The city has a glorious past since it has been for over two millennia the privileged stronghold of Byzantines, Arabs, and Normans. Rich in churches, valuable palaces and eighteenth-century villas, in 2002 it was awarded the title of World Heritage Site by UNESCO.

In spite of its small size (approximately 38,000 inhabitants), due to its geographical position, the city has been a hub for a area located in the neighbor plains of Gela and Catania. Such area overall counts about 325,000 inhabitants. It is famous for its ceramics handcrafts, an industry that has been flourishing since the era of the ancient Greeks.

A strategic plan was started in 2003, to identify and frame the factors impacting on the local area's performance. To draw up the plan, a team of consultants supported the Municipality of Caltagirone. Eighteen months were spent to collect data, to identify main stakeholders in and outside the region and conduct interviews with them, and to support the local administrators in outlining realistic goals and actions.

The plan defined a set of "visions" for the future of the city and the surrounding area. Such visions were then structured around a system of strategic goals, strictly intertwined each other. The guidelines for development (missions) were listed as follows: (a) develop and disseminate a strong culture of quality; (b) enhance the history, traditions, environmental resources of the area; (c) embellish/make the city more welcoming; (d) internationalize culture and trade relations; (e) adapt services to international standards; (f) focus policies on an area wider than the city, as a basis to increase the value provided by Caltagirone; (g) extend production chains; (h) combine innovation and tradition, but also art and technology (Bianchi and Tomaselli 2015).

The overall logic pursued by the plan was essentially based on framing the "drivers" of such development. For instance: which role would have played an improvement of key production chains, of local product and service quality, so to make the area more attractive to outside investments? This analysis would have involved the companies of the area in new projects aimed to generate a stronger business culture and to foster the development of professional services to local enterprises.

Ceramics plays an important role in the economy of the area. The city is populated by a myriad of shops exposing the typical products of this ancient art (dishes, jars, vases, tiles, candlesticks) engaging almost two hundred artisans.

After a period of strong growth, the industry entered into a stagnation phase, due to a loss of quality and the failure in research on new techniques and materials. Many competencies and professional skills related to the creation of the forms of handcrafts had been lost.

The plan described the critical issues about the development of this industry in the area. It was envisaged that most local artisans were dedicated mainly to the production of traditional and low quality products. Main customers were occasional travelers, tourists and a few residents.

The sector also produced floor and wall tiles for kitchens and bathrooms with traditional design. This market segment required a set of key-competences, closer to those of 'mass' than to 'handcraft' production, and therefore needed a proper management of commercial issues and manufacturing techniques. In spite of these requirements, local producers were still far from making a major shift from a mass-market production to the use of advanced manufacturing and commercial techniques.

The State Institute of Arts was experiencing difficulties in interacting with local producers.

The 'controlled designation of origin' recognition, which was attributed to Caltagirone's ceramics since the year 1993, never played the role of a driving force for the development of quality, production reliability and innovation. The launch of the Municipal Controlled Designation of Origin (DE.CO.P.) brand, by the Municipality of Caltagirone, in order to support and protect heritage, knowledge and experience of the ceramics handcrafting was still in its embryonic state.

It was necessary to recover the past values of art, and consistently connect them to those related to design.

In order to face such challenges, the strategic plan identified three lines of action, i.e.:

1. Pursuing a stronger coordination between actors. A new company was started to promote the development of good practices along the supply chain and to support local firms to refocus their design and commercial strategies, to better understand market structure and needs. The company's shareholders were the Municipal administration, the government of Sicily, the local agency for development, a local bank, the Institute of Art, and local businesses in the ceramics industry.
2. Improving the artistic and innovation skills of local artisan firms. A deep strategic reorientation of the Public School of ceramics, to restore its ability to support the development of local firms, through the training of craftsmen and professionals was needed.
3. Creating favorable conditions to improve product innovation and a marketing effort to gain a good international reputation. This implied the need to support startup firms in the ceramics industry in identifying their market segments, choose marketing channels (e.g., antiques dealers, jewelers, retail chains of high-class décor or design, high class hotels for higher quality products) and focusing other key players in the industry value chain (e.g.: designers, engineers, architects).

Although the efforts produced to draw up the plan were substantial, and the goals and lines of action appeared to be consistent with the described challenges, the implementation of the plan was poor. A weak entrepreneurial culture, fragmentation of initiatives, lack of coordination among stakeholders, and cultural resistance towards a local area perspective in policy making were major causes of such failure (Bianchi and Tomaselli 2015).

However, though such causes may provide a clear explanation of the cultural and socio-political difficulties inhibiting change, the main cause of the failure in implementing the described strategy can be attributed perhaps to the static and non-systemic view adopted by planners.

Though single strategic resources (e.g. infrastructures, knowledge, businesses, cultural heritage) had been explicitly considered in policy design, the plan was not able to capture the effects that adopted policies might have generated on their accumulation and depletion processes, over time, according to alternative scenarios.

Also, the plan could not capture delays between causes and effects. For instance, it did not take into consideration the time that the designed policies to improve R&D and artistic/innovation skills would have required in order to generate the expected effects on ceramics quality. Furthermore, it did not consider the delays through which the planned investments to improve the quality of the city's museums might have generated an improvement in the area's image (Bianchi and Tomaselli 2015).

Such a static perspective underlies a bounded planning view, which is quite far from a policy outcomes evaluation. This implies a risk of inversion between means and ends. According to such view, building a strategic resource (such as knowledge, infrastructure, R&D capacity, a controlled designation of origin recognition) is implicitly considered as a goal to achieve, rather than a pre-condition to gain, in order to carry out effective policies aimed to affect the area's performance, in terms of both drivers and end-results (i.e., outcomes such as: employment and investment rates, or change in citizens' quality of life).

Furthermore, although the plan mentioned possible trade-offs and/or synergies in space, e.g. related to different industries (such as: ceramics vs. tourism, vs. agriculture) or to market segments (such as artistic vs. industrial), there was not an attempt to measure the effects of planned policies on performance drivers and end-results.

All these weaknesses in the adopted approach to local strategic planning might have been overcome by using a DPM approach.

An example can be referred to the trade-offs between innovation and tradition in ceramic handcrafting. For instance, in the short run, focusing policies on tradition and continuity with the historical roots of the area is likely to consolidate the product image and its identification with the geographic area. This would increase the value (e.g. in terms of sales turnover, profits, or new jobs) generated by the product, and might foster further efforts oriented to focus the local area's policies on tradition's preservation (reinforcing loop "R1" in Fig. 4.28). However, focusing policies on only this direction might generate product obsolescence in the long run. In fact, new market trends might require a gradual adaptation of the product characteristics to the evolving values and needs of new generations of customers. A misperception of this need would increase the perceived product age, which would reduce the product fit with customers' evolving needs. This would decrease the value generated by the ceramics industry (balancing loop "B1" in Fig. 4.28).

On the other hand, in the short run, an aggressive innovation policy (e.g. aimed to foster an hybridization of ceramic crafts to embody emerging artistic traits from other cultures) could allow the product/local area to stay in an early maturity stage of its lifecycle and to increase the generated value (reinforcing loop "R2" in Fig. 4.28). However, in the long run an excess of focus on innovation might undermine the product identification with its area; this would reduce the value created by the industry to the benefit of the local area (balancing loop "B2" in Fig. 4.28).

Therefore, a proper mix between the two sets of policies might trigger a path to sustainable development. Actors must identify and analyze main feedback loops between variables affecting the described system behavior. It also implies a proper

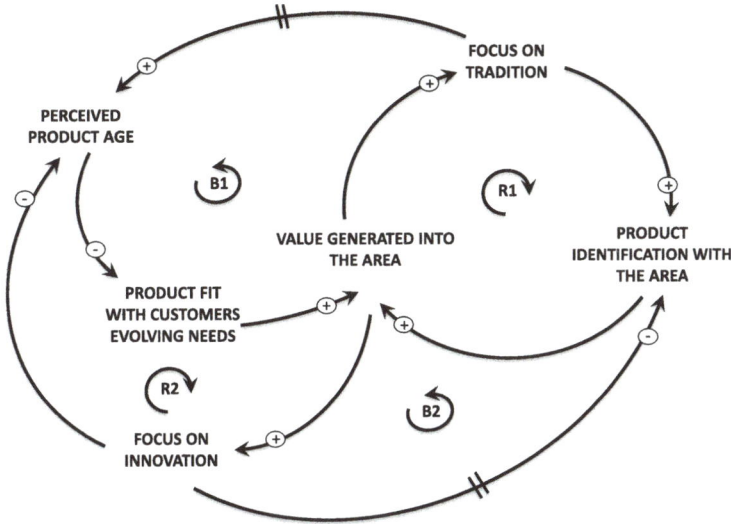

VARIABLE NAME	UNIT OF MEASURE	PERFORMANCE MEASURE TYPE
Perceived product age	Average age of product portfolio	2nd level performance driver
Product fit with customers evolving needs	% Customer needs satisfied by product features Number of product forms and used raw materials	1st level performance drivers
Product identification with local area	% Products telling the history of Caltagirone	1st level performance driver
Value generated into the local area	Cash flows, Profits, Employment rates, …	End-results

Fig. 4.28 Main feedback loops and performance measures associated with tradition versus innovation policies

methodological effort to identify and measure the performance variables related to the designed policies.

For instance, in the example portrayed in Fig. 4.28, an effort is made to define for each variable in the feedback loop diagram one or more corresponding performance measures. Regarding such measures, a causal analysis is also done: end-results are affected by first level performance drivers, which are—in turn—affected by second level performance drivers. In this case, the average age of product portfolio is the second level performance driver, which is influenced by an aging chain of stocks depicting the number of ceramic models or forms in the area's "portfolio" (strategic resources) at each lifecycle stage (Fig. 4.29). An aggressive product innovation strategy implies an increasing percentage of the new products stock on the total. Such state of the system can be pursued by alternative policies. In the Caltagirone strategic plan, the adopted policies were focused on a new company startup to foster R&D and on the strengthening of efforts to revitalize the School of ceramics. As remarked, the implementation of the two policies should not be evaluated in terms of simple output measures (i.e. number of started or accomplished projects), but especially in relation to their outcomes. Such outcomes can be detected based on the identification and measurement of the indirect policy effects on the identified performance drivers and end-results. A dynamic performance management synthetic view of the described context is depicted in Fig. 4.29.

For simplicity, we consider here as an end-result only the financial value generated into the geographic area by the ceramics industry. This can be referred as a synthetic expression of the income or cash flows earned in a given time span (e.g. a year) by companies in the area. On the one side, the accumulation over time of this value contributes to increase the stocks of equity and financial resources of local companies. On the other side, it provides a basis for further investments in the area. Such investments are here depicted as an accumulation into two different synthetic

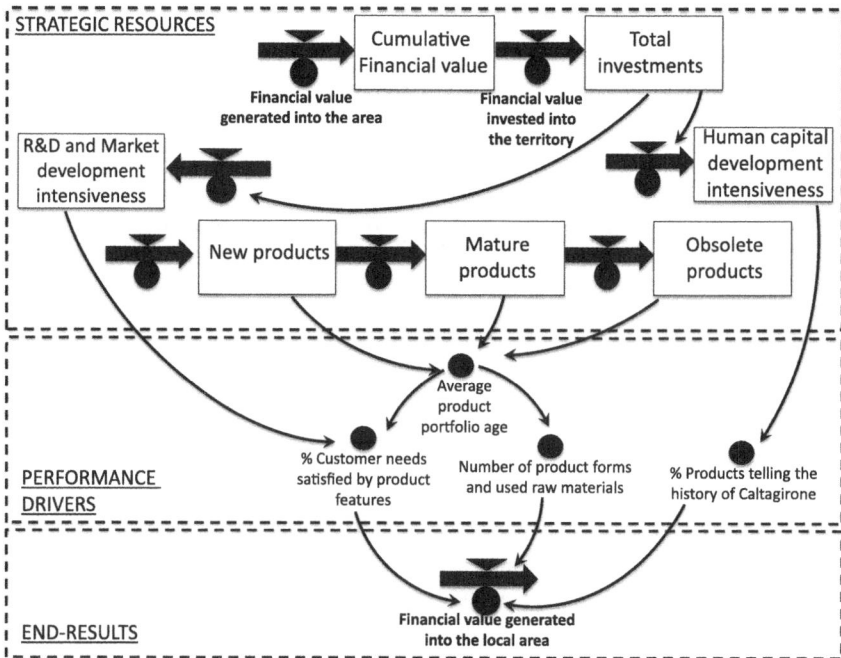

Fig. 4.29 A DPM chart framing the tradition versus innovation trade-off

strategic resources, i.e.: "R&D and Market development intensiveness" and "Human Capital Development intensiveness". The two stocks are an expression of the quality and volume of investments done in the area to ensure that the development activities focused on innovation and tradition will effectively impact on the key-drivers affecting end-results.

Overall, an effect of the static and non-systemic view that was adopted in sketching the Caltagirone strategic plan was a lack of attention on the role that an individualistic culture, a fragmentation of jurisdictions and administrative processes encompassing different public and private organizations in the area would have affected in delaying and tackling the implementation of the change process designed by the plan.

In the short term, a more realistic strategy might have been focused on the pursuit of a gradual improvement in key-actors' culture and knowledge, as well as in organizational management processes (i.e. R&D, production, commercial skills) along the ceramics value chain. A more open and collaborative culture and a stronger key-actors' knowledge of the historical roots of Caltagirone might have encouraged the start of partnerships on innovating projects in the field of ceramics, e.g. to feed the use of new materials in production processes (performance drivers), leading to the launch of new product lines, the search of new market segments, leading to an increased value generated in the area—e.g.: sales volumes and revenues, profits, cash flows, employment rates, new companies (end-results). In the long run, such results would have significantly strengthened the socio-economic structure of the area and therefore would have increased its relative attractiveness towards an improvement in the net of partnerships, both on the international markets and on other adjoining industries.

Figure 4.30 shows how such long-term effects of policies, aimed to combine tradition and innovation in the ceramics industry, may improve the local area's attractiveness (performance driver). This would determine an increase in the stock of companies operating in the area in the ceramics and other adjoining industries (end-result). A higher number of companies located in a stimulating and competitive geographic area would contribute—other things being equal—to increase the intensiveness and inclination of players to network (performance driver). This would, on one side, further increase the area's attractiveness, which might also positively affect the population rate. On the other side, it would also increase the number of partnerships and the employment rate (end-results). Such effects might further amplify the area's growth rate, since a higher stock of skilled employees in the area (strategic resource) would make the area more attractive to potential investors. Also, a higher stock of companies located in the area might further increase the intensiveness and scope of collaboration projects.

The capability of players in the area to frame and affect the driving forces of such growth is a fundamental condition to ensure sustainable development. For instance, limits to growth might gradually originate from an increasing population, leading to saturation in the provision processes of different services in the area (e.g. housing,

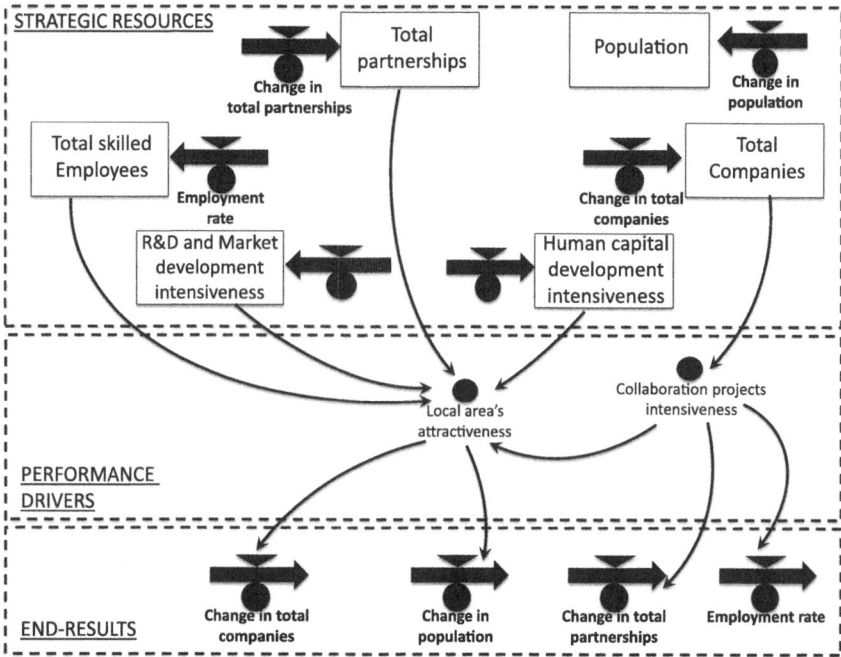

Fig. 4.30 DPM chart framing long-term effects of tradition versus innovation policies on local area's attractiveness and on collaboration projects intensiveness

health care, police, education, traffic). Ignoring such limits to growth might generate a shift from a development to a crisis pattern in the management of the local area.

The analysis shows that a DPM view can support local strategic planning by allowing decision makers to discern short from long-term policies; by supporting the linkage of strategic goals to appropriate measures aimed at gauging expected and emerging results. It also helps decision makers to clearly distinguish means from ends, and to indentify different "layers" of performance measures, starting from the identification of end-results and of corresponding sequentially related performance drivers. Overall, this approach may support decision makers to better conceptualize the relevant system related to the investigated problem behavior, and to better implement designed strategies.

References

Ammons, D. (2000). Benchmarking as a performance management tool: Experiences among municipalities in North Carolina. *Journal of Public Budgeting, Accounting & Financial Management, 12*, 106–124.

Begg, I. (1999). Cities and competitiveness. *Urban Studies, 36*(5–6), 795–809.

Behn, R. D. (2003). Why measure performance? Different purposes require different measures. *Public Administration Review, 63*(5), 586–606.

Behn, R. D. (2011). Be aware (and beware) of the Campbell's Law. *Behn Report, 9*. Available at http://www.hks.harvard.edu/thebehnreport/All%20Issues/August2011.pdf

Bianchi, C. (2010). Improving performance and fostering accountability in the public sector through system dynamic modeling: From and 'external' to an 'internal' perspective. *Systems Research and Behavioral Science, 27*, 361–384.

Bianchi, C., Bivona, E., Landi, T., & Ricci P. (2010). Fostering accountability in public utilities: The "ACQUA SPA" case-study. In Bianchi C. et al. (Eds.), Applying system dynamics to foster organizational change, accountability and performance in the public sector: A case-based Italian perspective, *Systems Research and Behavioral Science*, Vol. 27, pp. 395–420.

Bianchi, C., & Cognata, A. (2010). Improving performance through a dynamic balanced scorecard for opera houses. In Bianchi C. et al. (Eds.), Applying system dynamics to foster organizational change, accountability and performance in the public sector: A case-based Italian perspective, *Systems Research and Behavioral Science*, Vol. 27, pp. 395–420.

Bianchi, C., & Montemaggiore, G. (2008). Enhancing strategy design and planning in public utilities through "dynamic" balanced scorecards: Insights from a project in a city water company. *System Dynamics Review, 24*(2), 175–213.

Bianchi, C., & Navarra, D. (2013). *Enhancing performance management and sustainable development through e-government policies in urban areas: A system dynamics approach.* ASPA Conference, New Orleans, March 15–19.

Bianchi, C., & Rivenbark, W. (2014). Performance management in local government: The application of system dynamics to promote data use. *International Journal of Public Administration, 37*, 945–954.

Bianchi, C., & Tomaselli, S. (2015). A dynamic performance management approach to support local strategic planning. *International Review of Public Administration, 20*(4), 370–385.

Bianchi, C., & Williams, D. (2015). Applying system dynamics modeling to foster a cause-and-effect perspective in dealing with behavioral distortions associated with a city's performance measurement programs. *Public Performance Management Review, 38*, 395–425.

Borgonovi, E. (1996). *Principi e sistemi aziendali per le Amministrazioni Pubbliche (Management principles and systems for public administrations).* Milano: Egea.

Bouckaert, G., Laegreid, P., & Van de Walle, S. (2005). Trust, quality measurement models, and value chain monitoring: Symposium introduction. *Public Performance and Management Review, 28*(4), 460–464.

CABE, Commission for Architecture and the Built Environment. (2013). http://webarchive. nationalarchives.gov.uk/20110118095356/http://www.cabe.org.uk/case-studies/hammarby-sjostad

Campbell, D. T. (1969). Reforms as experiments. *American Psychologist, 24*(4), 409–429.

Cavenago, D., & Trivellato, B. (2010). Organizing strategic spatial planning: Experiences from Italian cities. *Space and Polity, 14*(2), 167–188.

Eterno, J. A., & Silverman, E. B. (2012). *The crime numbers game: Management by manipulation.* New York: CRC Press/Taylor & Francis.

Forrester, J. W. (1969). *Urban Dynamics.* Pegasus: Waltham.

Frey, B. S., & Pommerehne, W. W. (1989). *Muses and markets.* Blackwell, London: Explorations in the Economics of the Arts.

Ghaffarzadegan, N., Lyneis, J., & Richardson, G. P. (2011). How small system dynamics models can help the public policy process. *System Dynamics Review, 27*(1), 22–44.

Gutiérrez, U. M., & Menozzi, A. (2008). Board composition and performance in state-owned enterprises: Evidence from the Italian public utilities sector. *European Financial Management Association Annual Meeting Proceedings*, Athens, June 25–28.

Heilbrun, J., & Gray, C. M. (2001). *The economics of art and culture.* Cambridge, UK: Cambridge University Press.

Holzer, M., & Kloby, K. (2005). Sustaining citizen-driven performance improvement: Models for adoption and issues of sustainability. *The Innovation Journal: The Public Sector Innovation Journal, 10*(1), 1–20.

Jessop, B., & Sum, N. L. (2000). An entrepreneurial city in action: Hong Kong's emerging strategies in and for (inter) urban competition. *Urban Studies, 37*(12), 2287–2313.

Kim, H., Mac Donald, R., & Andersen, D. (2013). Simulation and managerial decision making: A double loop learning framework. *Public Administration Review, 73*(2), 291–300.

Lever, W. (1999). Competitive cities in Europe. *Urban Studies, 36*(5–6), 1029–1044.

Lind, M., Ostberg, O., & Johannisson, O. (2009). Acting out the swedish e-government action plan: Mind and mend the gaps. *International Journal of Public Information Systems, 5*(2), 37–60.

Meyer, M. W., & Gupta, V. (1994). The performance paradox. *Research in Organizational Behavior, 16*, 309–369.

OECD. (2000). The reform of metropolitan governance, http://78.41.128.130/dataoecd/3/17/1918016.pdf

OECD. (2001). *Cities for citizens. Improving metropolitan governance*. Paris: OECD Publications Service.

Porter, M. (1995). The competitive advantage of the inner city. *Harvard Business Review, 73*(3), 55–71.

Razumeyko, N. (2011). Strategic planning practices in North-West Russia. European influences, challenges and future perspectives. In Adams N., Cotella G., & Nunes R.(Eds.) Territorial Development, Cohesion and Spatial Planning: Building on EU Enlargement (pp. 402–421), New York, NY: Routledge.

Rivenbark, W. (2001). *A guide to the North Carolina local government performance measurement project*. Chapel Hill, NC: Institute of Government.

Rivenbark, W., Ammons, D., & Roenigk, D. (2005). *Benchmarking for results*. Chapel Hill, NC: School of Government.

Rivenbark, W., & Peterson, E. (2008). A balanced approach to implementing the balanced scorecard. *Popular Government*, Fall, 31–37.

Rivenbark, W., & Pizzarella, C. (2002). *Final report on city services for fiscal year 2000–2001*. Chapel Hill, NC: Institute of Government.

Van Dooren, W., Halligan, J., & Bouckaert, G. (2010). Performance management in the public sector. New York: Routledge.

Van Thiel, S., & Leeuw, F. L. (2002). The performance paradox in the public sector. *Public Performance & Management Review, 25*(3), 267–281.

Chapter 5
Applying Dynamic Performance
Management to Enterprises

5.1 Introduction

This chapter will illustrate two examples of DPM applied to enterprises. Cases will show how DPM can support a learning-oriented approach into the P&C processes of business organizations.

From the examples that will be discussed, the reader will learn how DPM can support business decision makers to frame the specific dynamic complexity of the industry where their companies operate and to pursue a sustainable organizational growth.

5.2 Applying DPM to Support Entrepreneurial Learning in Business Startup and Restructuring Processes

In this section, a *microworld* using a DPM approach, to support strategic learning in business startup and restructuring processes will be illustrated.

Microworlds (otherwise stated as: "interactive learning environments", or "management flight simulators") are SD-based simulation models aiming to foster policy debate. The use of such simulators, supported by a learning facilitator, can help policy makers understand the dynamic relationships between strategic resources and performance variables. In fact, the elicitation of the causal chain between them may enhance decision makers' learning processes and, thus, their ability to comprehend how different strategies might affect organization results over time. In particular, such simulators offer managers a "virtual world" where they can test their hypotheses and evaluate the possible effects of their strategies without bearing the costs and risks of experimenting with them in the real world (Morecroft 1987). Virtual worlds are effective if they enable learners in practicing "reflective thought", i.e. an ability to exercise a reflective conversation with the situation.

© Springer International Publishing Switzerland 2016
C. Bianchi, *Dynamic Performance Management*, System Dynamics
for Performance Management 1, DOI 10.1007/978-3-319-31845-5_5

When *microworlds* are used, this occurs in the mode of design, i.e. through a "transitional object" (that is a model) which is able to act as a tangible aid to imagination and learning, and to help people make better sense of a partly understood world (Morecroft 2007, p. 374).

When "flight simulators" are used to support strategic learning processes of a management team, each individual carries around an image of the organization taken from the real world and filtered through his or her own mental models. Such images of the real world may vary according to the experience, organizational role, ambitions, and objectives of each person. Through the use of a simulator—as a "transitional object" providing a *virtual world* to play with—a learning facilitator may support a group of policy makers to elicit and compare their own mental models with the goal to: (1) improve the quality of their understanding of a dynamic and complex world, (2) gain a common shared view of why, how, and when performance is affected, and (3) improve the quality of their decisions.

The role and value of SD *microworlds*, in terms of supporting learning through participating in simulated business management, has been widely described in the literature (Morecroft and Sterman 1994; Davidsen et al. 2000). Romme (2004) asserts that mature students (for example on MBAs), who may have significant business experience, are more likely to benefit from microworlds than less mature ones. Mohapatra and Saha (1996) describe simulator trials that they argue corroborate their hypothesis that learning and performance are positively correlated.

However, whilst the potential benefits of simulator use might be established as a general principle, there are practical issues of validity and ease of use that relate to an individual simulator such as the one described here, i.e.:

- In the development of any business enterprise simulator, its designers use theories and business fundamentals drawn from their own knowledge and experience. Of course, because the simulation designers must choose specific procedures and theories, their personal bias can become a factor and—as Goosen et al. (2001) and Grossler (2004) suggest—the learning benefits of a specific simulation can thereby be affected by what is and is not chosen as the knowledge base, which mostly depends on both the bounded rationality of the learner and of the modeler/learning facilitator.
- The usefulness of any simulator to an entrepreneur (as opposed to a student) will be a function of how easy he or she finds it to use and how realistically it portrays his or her own business. Simulator developers typically look to prove their simulators with representative user groups.[1]

---

[1]For example, Rouwette et al. (1998) proved their management game to introduce employees of housing associations to new market conditions. Ford (1996) has described how his Snake River water resources model was perceived by a group of 30 individuals gathered together. Both these simulators, of course, reflected specific cases, either a particular industry sector or one actual system, although Winch and Arthur (2002) have demonstrated that at the 'insight' level generic models, possible parameterized with general values representing an individual company, can be acceptable by users as sufficiently suggestive of their own actual firm.

– The effectiveness of a SD-based microworld as a learning aid is subordinated to the consistency of the adopted teaching strategies with the goal of fostering a "double loop" learning process in a group of learners using a simulator. This implies that a simulator is adopted as a vehicle to develop the learners' ability to perceive and map the hidden "structure and behavior" feedback relationships underlying business performance. To this end, simulators are generally used as a teaching support to a case-study. After discussing the case's debating points, the learners are engaged in a group simulation session, by using a microworld. This puts the learners in a learning and 'risk safe' environment, where they can experiment an organization and a decision context having the same features described in the case. Proper use of a SD microworld implies that a facilitator supports the learning process in both group and de-briefing plenary sessions. Such approach does not primarily aim to generate a competition between groups, with the goal to maximize or minimize a set of targets—as it happens when business games are used. The emphasis of simulation, teamwork, and plenary de-briefing sessions in this context is, rather, on supporting the acquisition by the learners of a feedback view to enhance DPM.

The simulations described in this section reflect the efforts of the microworld designer to ensure that it adequately reflects the reality of a small firm's operations and that it can produce a realistic and believable portrayal of how startup, business restructuring and growth could arise, and of the possible scenarios for enabling firms to move into long term sustainable growth.

5.2.1 Applying DPM to Support Entrepreneurial Learning in Business Startup and Restructuring Processes (Continued): CompuGames Dynamic Business Plan Simulator

In this section the "CompuGames" case-study and simulator will be illustrated, with the goal to show how the use of SD microworlds may support small business startup and restructuring planning processes, according to a DPM approach.

"CompuGames" is a small business that was started about two years ago. It operates in a market niche in the videogames and entertainment industry, on a boundary between the real and the virtual world. Based on the multi-year experiences of its two founders, the company is positioned in the industry as an independent videogame developer. Its products are positioned on a medium-high price level. Currently, after less than two years from its startup, its operations may only rely on the capabilities and creativity of its two founders, who are supported by three employees, i.e.: a designer, an assembler, and a sales agent.

Recently the company has adopted a new technology that enhances the so-called "augmented reality" (AR): through AR visors and software, virtual three-dimensional objects can be visualized in a very realistic manner in a virtual environment, so that the users can interact with them. Such technology can be used in other industries, e.g.: construction industry simulation, medicine, cinema, archeology, and industrial prototypes. This has allowed CompuGames to enter new adjacent market segments, in respect to the original one.

Also, CompuGames has designed a transportable and easy to install projection system, which is suitable to be carried to exhibitions and special events. Another application the company is now developing is related to "life-size videogames".

5.2.2 The Industry Where CompuGames Operates

The videogames industry is one of the fastest-growing sectors in the world: the USA, Japan, and Europe are the most developed markets. According to the Year Book 2003/2004 by ISFE (Interactive Software Federation of Europe), if one only takes into account software sales, the videogame industry has reached a value of Euro 15.3 billions worldwide. If one also considers hardware, sales turnover reaches Euro 20 billions.

A home videogame consists of four main components: (1) console, (2) audio/video output; (3) peripheral control devices, and (4) software. The console is a kind of computer that uses a microprocessor, optimized for enhanced graphics. It receives information from peripheral control devices, then it processes them, and —based on the software game instructions—it transmits signals to a display through an audio/video system.

Five main players are in the industry, i.e.: (1) content providers, (2) software developers, (3) publishers, (4) console producers, and (5) retailers.

Content providers play a role when the videogame is not based on a completely new idea, but is instead inspired to an existing product or other commercial idea. The owner of the intellectual property of the topic on which a videogame can be based may receive a royalty that approaches the 10 % of the product retail sales revenues.

Software developers (such as CompuGames) are those firms that actually produce the videogame. There can be three different kinds of software developers, depending on which their ownership is held by: (1) a software publisher; (2) a console producer; (3) an independent developer. Only a few independent developers are able to self-finance their own software development projects. For this reason they are often forced to look for financial support from software publishers or console developers.

In this industry it is crucial to keep the Research & Development cycle time as short as possible: on average around 6 months. Time to market and return on invested capital are critical factors to succeed.

The product lifecycle is short: more than 50 % and sometimes up to the 80 % of sales revenues is earned within the first year from the introduction of a new product into the market.

Videogame publishers are used to finance new product ideas. Such companies are usually larger than software developers. Main critical issues they must face to manage their competitive position are related to marketing and to the management of their relationships with console producers.

Promotional effort must start six months before the introduction of a new product on the market. It may cost up to 5 times more than product development.

Console producers manage three main areas: (1) design, production and marketing of the hardware on which the software game is based; (2) videogame software development and publication, and (3) relationships with retailers. Most console producers are large firms, whose competitive advantage is mostly based on their capability to periodically generate a technological discontinuity, through innovation, so to strengthen market entry barriers. For such companies, getting a large market share through high promotional investments is an important means to gain a sustainable competitive advantage.

Retailers are located in the final stage of the industry supply chain. Hardware and software videogame products are sold through different channels; the most important are specialized shops and toy stores.

The analysis developed so far gives an idea of the challenges small independent developer firms must face, particularly if they aim to publish their products with their own branded logo.

5.2.3 The Dynamic Business Plan Simulator

The two founders of CompuGames were puzzled by the challenges imposed by the company's competitive system. They wondered whether in the next three-four years their business would have been able to face and counteract the industry challenges, and to develop a sustainable competitive advantage in the new market segment, so to build on the knowledge base developed in the years before the company startup.

With the aim to find an answer to such question, the company founders decided to sketch a dynamic business plan. Such plan could have supported them to frame and overcome the risks for the company, in the next four years, to either: (1) become a marginal player in the industry, due to a lack of resources that could support new product development and sales growth, or (2) undertake a too fast and unsustainable growth rate, which might transform it into a 'giant with feet of clay'.

A business plan simulator based on a DPM perspective was built.

The model consists of five main subsystems: Commercial, Research & Development, Production/Assembling, Internal support processes, and Financial.

5.2.3.1   The Commercial Subsystem

If one takes a 4-year simulation time horizon, the product lifecycle of CompuGames can be referred not only to the existing 4 videogames the company has on the market, but also to the new products it will be able to launch by adopting the new technology based on holographic displays.

The product lifecycle is on average 2.5 years long. Once designed, a videogame stays in an 'introduction' stage for 6 months. Then, it enters in a 'growth' stage for about 7.5 months. Afterwards, it reaches a 'maturity' stage where it stays for about one year. In the next 6 months, the product enters in a 'decline' stage, before its dismissal from the company portfolio (Fig. 5.1).

Main critical success factors in the market are related to those variables affecting business image, i.e.: (1) product "quality/price" ratio, (2) delivery delay to retailers (normal time is three weeks), (3) promotional spend (mainly through TV spots, advertising on newspapers and magazines, and the participation to fairs), and (4) commercial support through sales agents.

The used component parts affect product quality: the company may either purchase medium or high quality component parts. Purchasing high quality parts implies a 20 % cost increase, in respect to the medium, which is about €9.5/piece.

A new product must be introduced on the market at a price that is the 80 % of its reference, which is referred to the maturity stage. When a product enters in a growth stage, then its price will be increased to the 90 % of the reference. For those products in a decline stage, price will drop again to the 80 % of the reference (Fig. 5.2).

The company's current average price is €250 per unit. Competitors are used to set a price that may range from a minimum of €200 to a maximum €350 per unit.

Price may affect average sales orders per customer. Demand elasticity to price may differ according to the product lifecycle stage. However, a low price strategy may allow the firm to gain up to 50 % more than the reference price, in terms of

Fig. 5.1 CompuGames
product lifecycle

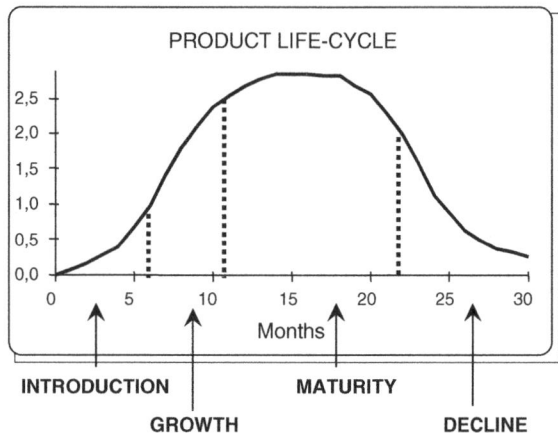

Fig. 5.2 Sale price as a function of product life-cycle stage (example related to a €250 reference price)

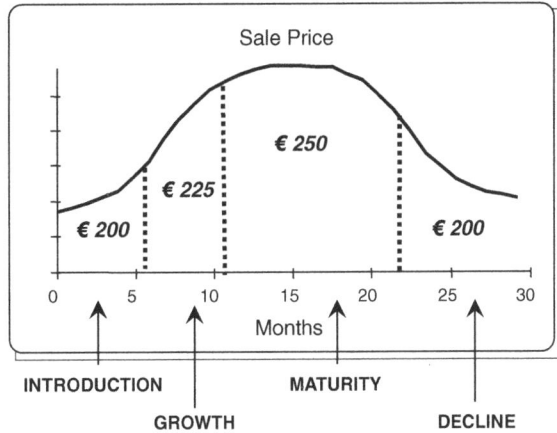

market demand. At the same time, a high price strategy may imply a drop in demand up to the 50 % of the reference.

Promotional spend is needed to sustain the product on the market: it may allow the company to increase its customer base and sales turnover.

The promotional spend in the market may range from a minimum of €50,000 and a maximum of €200,000 per quarter. The business currently adopts a €100,000 promotional spend per quarter, to support its own 4 products, which are on a "late maturity" stage, with an 8 % market share. According to market research and the personal experience of the business owners, this budget would allow the firm—other conditions being equal—to gain on average around 150 new customers (i.e. retailers) in a quarter: however, the yield of promotional spend would significantly decrease with the acquisition of a higher market share. A 40 % market share would drop the contribution of current promotional spend to 70 new customers/quarter. This yield would further drop to around 5 customers/quarter, if the company market share would approximate to 60 %.

The yield of promotional spend is also affected by the product portfolio age. In fact, when the product portfolio's average age is younger than the maturity stage, the needed promotional spend to sustain the customer acquisition rate may increase up to 30–35 % more than the normal level. For example, a product portfolio characterized by an average "late growth" stage (corresponding to about a 1-year product average age) would imply a reduction of the yield from promotional spend of 10 % or more, in respect to its potential. Though this reduction may not seem dramatic, if observed in relative terms, it is remarkable if one considers that the high market competitiveness implies an 8 % normal customer loss rate per week (on the company customer base) and that the net change in the customer base can be also negatively affected by: (1) a low or dropping quality/price ratio, (2) a too narrow product portfolio (i.e., less than 4 products), (3) an early growth or late mature/old product portfolio, and (4) a delivery delay higher than the reference level.

While innovating the product portfolio is a vital strategy to pursue a competitive advantage in this market (provided that product lifecycle is short), this implies a number of financial challenges for commercial strategies. In fact, a 'young' product portfolio both requires a lower sale price and a higher promotional spend, in respect to maturity. In principle, increasing the average product reference price could indirectly fund aggressive new product launch policies. However, this would reduce (other things being equal) the "quality/price" ratio, which would decrease market demand.

A fourth commercial policy lever is sales agent support. Hiring salespeople may affect new customers acquisition and sales turnover rates in a similar way to what has been commented regarding the promotional spend. In fact, on average each agent can contribute—other conditions being equal—to gain up to 12 customers per quarter. However, the returns from a sales agents acquisition policy might diminish due to: (1) an increasing company market share, and (2) a product delivery delay higher than the reference.

For instance, a 20 % market share may reduce the normal returns—in terms of customer acquisition rates—to one half of the reference level that can be reached when the company market share is minimal. Such returns would drop to a 10 % when the company market share approaches a 50 %, and would further diminish with an increasing market share.

Similarly, a 6-months perceived product delivery delay by customers (i.e. twice the reference level) would imply a loss of about a 20 % in the capability of sales agents to get new customers (in respect to the normal). An 8-months delivery delay would imply a loss of about 60 % in such returns, which would further drop if the delivery delay would increase more.

CompuGames currently has only one sales agent. His gross salary is €4000/month. To this cost, a 3 % commission on sales revenues must be paid to him. Hiring new sales agents would require a 6-weeks time before they can be productive, after training. In the industry, the average time to quit the firm by sales agents is 4-years.

Also, word-of-mouth may contribute to customer acquisition. On average, for each satisfied company customer, CompuGames may get a new customer (i.e. retailer) every 5 months.

5.2.3.2 The R&D Subsystem

Research & Development is vital to launch new computer games.

R&D processes are carried out through three main steps, i.e.: (1) design, (2) engineering, and (3) final product configuration. Quality control is also carried out to support the first two steps of the process. R&D defects may rise due to an excess workload on the available staff (designers). Also, the capability of the firm to detect defects in R&D processes is affected by workload. Excess in using overtime working tends to increase the number of defects in R&D. This may lead to an increase of the: (1) time to launch new products on the market, in respect to the reference R&D time (i.e. 6 months), and (2) percentage of undetected defects in the R&D processes, leading to a higher probability of failure in new product manufacturing.

An increase in the time to launch new products would contribute to increase (other things being equal) the product portfolio average age. This would reduce the effectiveness of commercial efforts, in terms of new customer acquisition, sales orders and revenues rates.

A decrease in the quality of design would imply an increase in the used component part costs up to four times of the standard level. For example, a 20 % reduction in design quality would increase of about 50 % the cost of component part; a 40 % reduction in design quality would double such costs. A 50 % reduction in design quality would even triple them.

The company currently employs three designers. Their unit gross salary is €5200/month. Overtime cost is €45/h. It is necessary to use overtime (up to a 10 h/staff/week) if the total available working hours are not enough to perform the workload.

It takes 3 months to train newly hired design staff. In the industry, the average time to quit the firm by design staff is 4 years.

5.2.3.3 The Production/Assembling Subsystem

The new models that are configured by the R&D processes can be launched on the market, and promoted through commercial policies. Sales orders backlogs are delivered through the production/assembling subsystem. The following resources affect production and deliveries: (1) inventory, (2) assembly staff, and (3) assembly machinery capacity.

Each product uses on average 10 component parts. Component parts are purchased based on expected demand. Desired inventory levels are based on the expected demand in the next three weeks. The minimum assembly time is 3 weeks.

The company currently employs one assembly staff unit. Her salary is €4400/month. Overtime cost is €35/h. A maximum of 10-h per staff overtime can be performed each week. Though overtime work may allow the firm to overcome peaks in product demand, a prolonged resort to such policy may generate employee burnout; in fact, overtime hour productivity may drop up to 60 % than normal (which is equal to 5 component parts/hour).

Hiring new assembling staff would require a 3-weeks time before they can be productive, after training. In the industry, the average time to quit the firm by sales agents is 4-years.

The current assembly machinery capacity is 3000-component parts/week (i.e. 300 product units/week).

The investment costs associated with an increase of such capacity of an extra 100 product units/week are €270,000. In order to prevent obsolescence, machinery must be replaced every two years.

5.2.3.4 The Financial Subsystem

Figure 5.3 shows the company balance sheet, referred to the beginning of the planning time horizon. It portrays a high percentage of fixed assets over the total net investments, and a quite high "debts-to-equity" ratio. This is due to the fact that the two owners-entrepreneurs started the business by only investing financial resources from their own family's assets and by exploiting available bank credits. The two business owning families have subscribed and invested all the company's equity.

CompuGames may rely on a €6.8 million maximum bank credit. Such credit may change as a function of: (1) the value of business equity perceived by banks, and (2) the investment turnover rate (i.e. the "sales revenues/net investments" ratio) perceived by banks.

The initial personal assets of the business owning families is €15 millions. Such stock is increased by periodical dividends withdrawals and salaries paid to business owners/managers; it is depleted by the current family expenses. Therefore, new equity investments from the business owning families' assets will be possible as a function of the existing family assets and of their inflows.

The two business founders receive a forfeit stipend of €3000/month (as members of the company Board), and a salary as designers.

An increase in the salaries paid to business owners/managers increases the business owning families' satisfaction level about the business. An increase in the salaries paid to business owners/managers will generate a gradual rise in the current business owning families' expenses. The founders' families will increase their dividend withdrawal requests, based on the perceived growth rate of the firm. Trust by the business-owning families (in terms of change in the satisfaction level) towards the company is an important factor in assessing CompuGames' stakeholder strategic performance.

In order to fund its financial needs, the company may also increase its equity through venture capitalist investments. Venture capitalists will be initially inclined to fund up to the 50 % of the prospected discounted cash flows at a 20 % interest rate on a perpetual annuity.

Venture capitalists expectations towards the company are affected by: (1) the business capital value, and (2) dividends paid to equity owners. Venture capitalists' trust (in terms of change in the satisfaction level) towards the company is another important factor in assessing CompuGames' stakeholder strategic performance.

Other financial parameters are: (1) terms of payment allowed to customers (four weeks), (2) terms of payment allowed by: (2-a) component part suppliers (3.5 weeks); (2-b) machinery suppliers (1-year); (2-c) promotional service providers (8 weeks), and (2-d) administrative service providers (24 weeks).

Regarding the specific governance and financial profile of CompuGames, as an example of a small business whose equity is fully owned by its founders, most critical issues on which the use of the simulator may enhance decision makers' learning processes are focused on challenging:

BALANCE SHEET (Euro)	Initial value
NET INVESTMENTS	**18.957.849**
Fixed assets:	**8.111.667**
- Assembly machinery	8.100.000
- Capitalised design costs	11.667
Working capital:	**646.352**
- End products inventory	112.500
- Component parts inventory	128.250
- Accounts receivable	405.602
Positive bank balances	**10.199.831**
EQUITY & LIABILITIES	**18.957.849**
Long term funds:	**14.919.648**
- Equity	6.819.648
invested by the family	*6.819.648*
invested by venture capitalists	*0*
- Long term bank debts	0
- Debts towards assembly machinery suppliers	8.100.000
Short term debts:	**4.038.201**
- Accounts payable	305.513
- for salaries	42.968
- for other industrial costs	305.220
- for promotional activities	3.200.000
- for administrative costs	184.500
- for taxes	0

Fig. 5.3 CompuGames initial balance sheet

1. A misperception by the business owners of the financial needs associated with the size of capacity and current investments that surviving in a highly competitive market may imply, and of the necessary resources that should (and might) be funded by drawing liquidity from the business-owning family assets,
2. A bias in profit and cash flow expectations leading to uncontrolled liquidity withdrawals from company bank accounts to satisfy family (current and non-current) needs, and
3. A lack of understanding by the business owners/founders of the business governance[2] implications that the subscription of the majority of the business equity by venture capitalists may imply.

The described phenomena are particularly frequent in unlimited liability companies, where owner/entrepreneurs and the members of their own families may misperceive the difference between business and personal assets.[3]

By using the simulator, learners may figure out relationships between the firm and the equity-owning family, particularly with respect to bias in profit and cash flows expectations leading to withdrawals by family members from the business bank accounts, to cover current expenses. Learning how to pursue a balanced growth, in compliance with both the business and family needs, is a concern for this simulator.

Balancing withdrawals and investments to achieve an adequate business-owning family and venture capitalists satisfaction that is compatible with business liquidity, and matching competitive policies with financial structure are the keys to survival and growth and continuity of both the business and its relationships with its financial stakeholders.

5.2.4   A Dynamic Performance Management View Portrayed by the Simulator

The interplay of the described business subsystems provides a sound basis for identifying the main performance and resource measures through which a dynamic business plan can support policy makers to assess the strategies that CompuGames might undertake in the future to foster sustainable development.

---

[2]An example of this might be related to the business owner's will to increase their own salaries or to appoint other family members as board members.

[3]Another important cause of uncontrolled liquidity withdrawals from company bank accounts can also be related to escalating behaviors originated by several family units owning business equity. This is especially the case of family businesses that have reached the second or further succession stages. Such firms are usually controlled by more than one family unit, as different brothers and/or daughters or even cousins may inherit the entrepreneurial function. In these circumstances, it may happen that deteriorating relationships between different equity-owning family units may give rise to escalating imitative withdrawals aimed at pursuing individual goals, to the prejudice of future business survival.

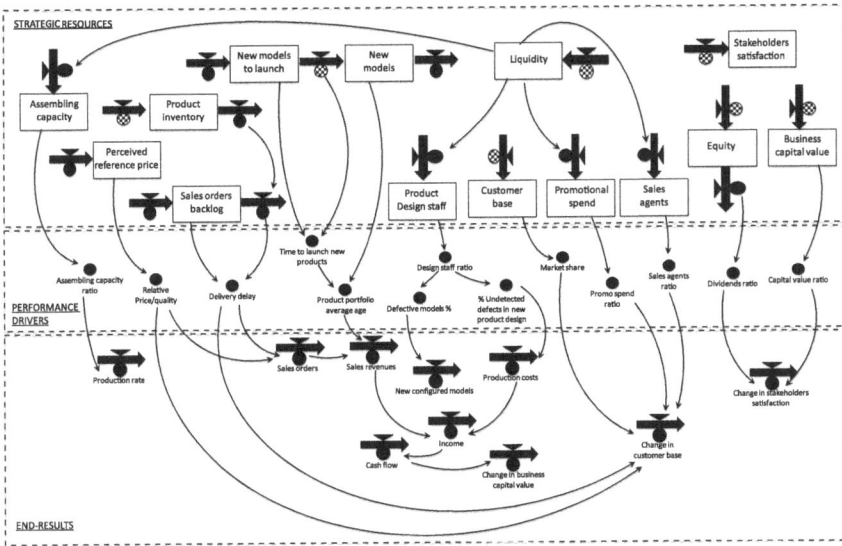

Fig. 5.4 A DPM view of "CompuGames" business plan

As shown in Fig. 5.4, which frames a simplified version of the CompuGames dynamic business plan simulator, both the end-results and performance drivers attain to the financial and competitive dimension of business success.

We will start to analyze Fig. 5.4 by identifying the performance drivers, in order to illustrate how company policies may affect them, and how they may in turn affect the end-results, which will feedback on the strategic resources generated by business routines.

Main critical success factors related to the commercial subsystem are: (1) "quality/price", (2) delivery delay, (3) promotional spend, and (4) sales agents support. Figure 5.4 shows how such factors can be framed (and measured) through performance drivers. The figure also shows how such drivers can be affected by leveraging strategic resources. It illustrates that commercial performance drivers affect sales orders and revenues, and the change in the customer base. More specifically, sales orders are affected by the "quality/price ratio" and product delivery delay. The change in the company customer base is affected by promotional spend, the stock of sales agents, the "quality/price" ratio, and product delivery delay. Market share is a fifth performance driver related to the commercial subsystem. It affects the change in the customer base: increasing the customer base and market share may generate diminishing returns from commercial investments (this implies the dominance of a balancing loop draining business growth).

If we now consider the R&D subsystem, performance drivers can be referred to: (1) the design staff ratio, (2) the percentage of defective R&D models, (3) the percentage of undetected defects in new product design, (4) the time to launch new products, and (5) the product portfolio age.

The design staff ratio measures the extent to which the company may rely on enough product design staff (strategic resource), in respect to desired level, which depends on the workload (i.e. to the stock of total models to launch). This performance driver affects both the percentage of defective models and the percentage of undetected defects. An "available staff/desired staff" ratio equal to or higher than one indicates an aptitude of R&D processes to keep both performance drivers low.

The higher the percentage of defective models is, the more rework will be necessary. This would reduce (other conditions being equal) the new configured models ready to be launched on the market (end-result), which accumulates in the stock of new models.

The percentage of undetected defects in new product design affects production costs in the long run. In fact, a lower product design quality implies higher risks of production failure and rework.

The ratio between the stock of total new models to launch and the new configured models in a given time defines the time to launch new products. Both the performance driver and the stock of new models on the market affect the product portfolio average age (a lower level performance driver). In fact, the less the time to launch new products, the younger the product portfolio average age will be. A too low or too high product portfolio age (corresponding to a high percentage of models in the early or late stages of the product life-cycle) reduces the sale price (in respect to the reference—i.e. maturity—level) and (other conditions being equal) sales revenues. Therefore, by keeping the product portfolio average age close to an early maturity stage and by launching in due time new products that may replace those which are approaching the obsolescence stage, the firm may effectively support the commercial subsystem through R&D.

The main performance driver related to the production/assembling subsystem is the assembling capacity ratio, i.e. a fraction between the actual and the desired capacity (in terms of both machinery and staff). An increase in this driver may foster an increase in the production rate (end-result), if the company holds enough component parts inventory and sales orders backlog. The production rate accumulates into the stock of product inventories (strategic resource). Holding enough product inventories may allow the company to feed the desired product-shipping rate. This will reduce the sales orders backlog and will affect the delivery delay.

Main performance drivers related to the financial subsystem are: (1) the dividends ratio, and (2) the capital value ratio. Such measures affect (and are indirectly affected by) the following end-results: (1) income, (2) cash flow, (3) change in the business capital value, and (4) change in stakeholders' satisfaction.

Income and cash flow affect equity and liquidity (strategic resources), respectively. Liquidity may feed the acquisition of physical and capacity resources. Equity may allow the business to distribute dividends. Cash flow also affects the change in the business capital value (end-results), which affects the capital value perceived by financial investors (strategic resource). Both distributed dividends and the business capital value (in respect to the desired levels) are performance drivers affecting the change in stakeholders' satisfaction (outcome end-result), which in turn affects the financial stakeholders' satisfaction level (strategic resource).

In the next section we will show an example of how using the CompuGames simulator in entrepreneurial development training programs may empower company owners/managers in developing business planning skills, based on DPM, and deal with dynamic complexity.

5.2.5 Playing with the Dynamic Business Plan Simulator

The CompuGames simulator is used in entrepreneurial training. The business case-study is first debated by participants in both team and a plenary de-briefing sessions, focused on the following debating points:

- How many new products (models) should CompuGames launch on the market (and when)?
- What strategic positioning should the company give to its products, in terms of: price, component parts quality, delivery delay, promotion, salesforce support?
- Should the company use different commercial policies according to the product portfolio average age? How?
- What strategic resources will the business have to build up (and when), in order to support its own commercial growth policies?
- When will the firm have to hire human resources to increase R&D and product assembling capacity? How many people should be hired?
- When will the firm have to increase its machinery assembling capacity? How many people should be hired?
- What effects will be generated on the company revenue and cost structure by an increase in the "product portfolio"?
- How would changes in the company's activity volumes and complexity level affect the cost structure? How fixed costs might change in the long run?
- How can one balance growth in the commercial subsystem with growth in the R&D, Assembling, and Financial subsystems?
- What financial needs would growth imply?
- Will it be possible (and realistic) to continue to fund such financial needs through the business-owners/founders family assets and/or bank credit?
- Would it be possible to raise equity capital from venture capitalists? What would such decision imply for the business, for its founders and the equity-owning families?
- What policies should the firm adopt in allocating internal flow of funds and cash flows? What are their implications on funding growth and satisfying financial stakeholders' expectations?
- What policies should the company adopt, in terms of stipends paid to their founders and other family members who might be hired to cover formal roles, e.g. in the Board or in key management areas?

– What effects would the subscription of equity by venture capitalists imply on the business founders' freedom to set dividends and to involvement of their own family members in key management areas?
– How will the dividend policy and the capital value of the firm affect financial stakeholders' satisfaction?
– How to balance business sustainable development with financial stakeholders' satisfaction?
– How will satisfaction level affect the overall company image?
– How to increase the capital value of the firm?

Based on the debating points, the case-study debriefing session has the main goal to help participants framing the dynamic complexity of the system and perceiving the need of a DPM approach to business planning.

After case debriefing, the different groups of participants are requested to sketch a 4 year dynamic business plan by using the CompuGames simulator. Team-based simulation sessions have the goal of pursuing a balanced achievement of four main sets of targets, i.e.: (1) business growth rate, (2) profitability and financial equilibrium, (3) stakeholders satisfaction (referred to both the business owning family and venture capitalists), and (4) business image (from both customers and stakeholders' satisfaction).

Through simulation, players can experience how complex it is to reach high performance levels at the same time, for multiple targets. For instance, they can perceive the trade-offs between business growth and stakeholders' satisfaction: distributing dividends may reduce company liquidity to the prejudice of business growth. However, reinvesting profits may reduce the level of trust by business owners and venture capitalist towards the firm capability to generate a payback on their own investments.

Similar trade-offs can be experienced regarding the relationships between business growth and profitability/financial equilibrium: if the product portfolio is in a mature stage, a low rate in new product and capacity investments may perhaps keep stable or improve business liquidity and profitability in the short term, but it may also prejudice business growth and financial results in the long run. On the other hand, a too high growth rate, in respect to the financial structure of the business, may prejudice solvency and profitability.

Such trade-offs are amplified by the complexity related to business startup, and to the small size of the company in a market requiring continuous R&D and commercial investments.

During the simulation sessions, learning facilitators support teamwork. Their role is to help participants perceiving and modeling the feedback structure associated with the system's simulated behavior over different simulation runs. In particular, participants are supported in framing the system causal structure according to a DPM perspective, as previously illustrated (Fig. 5.4).

The dynamic business plan simulator based on the CompuGames case-study consists of a start window (Fig. 5.5a) through which the players may have access to three sectors, i.e.: (1) an "input parameters" window (Fig. 5.5b), (2) a 'transparent

(a)

(b)

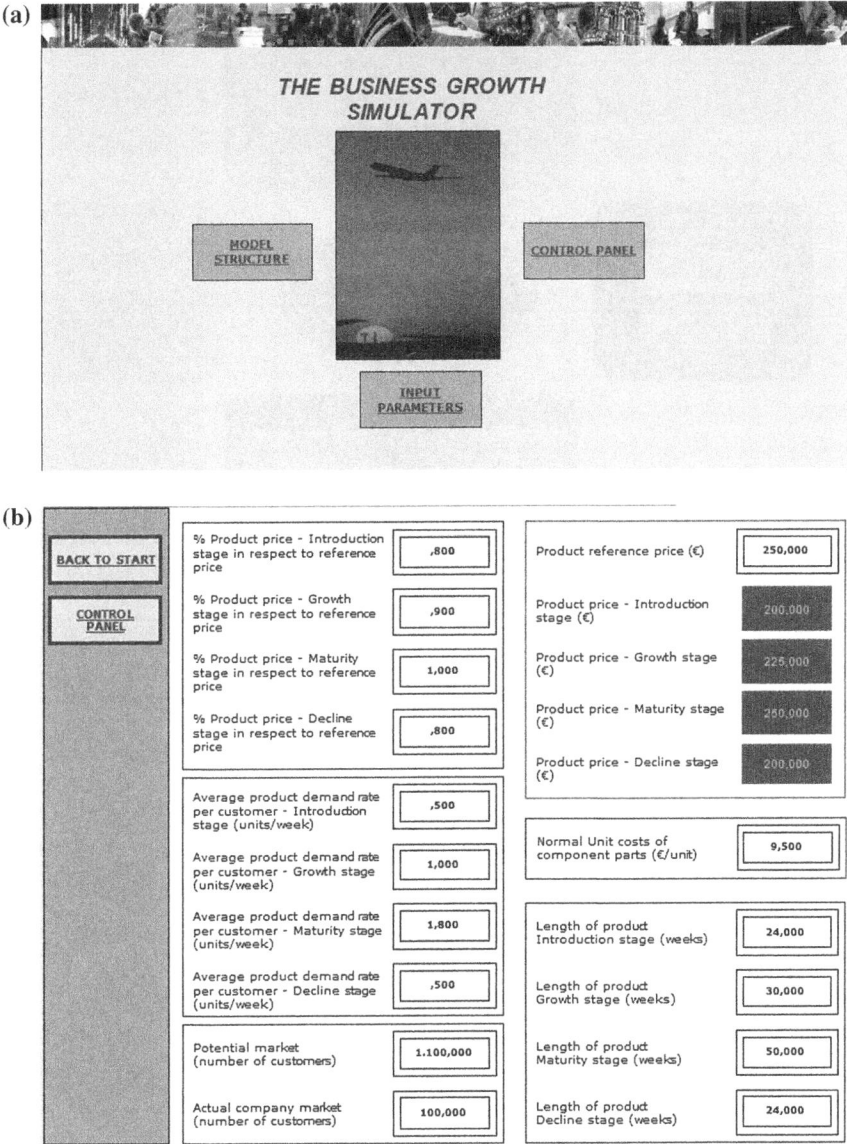

Fig. 5.5 **a** Start window. **b** Input parameters window. **c** Transparent box stock-and-flow model structure window. **d** Control panel

box' stock-and-flow model structure window (Fig. 5.5c), and (3) a control panel (Fig. 5.5d).

The "input parameters" window allows one to set initial values for a number of structural factors affecting performance, i.e.: market size (in terms of customers),

(c)

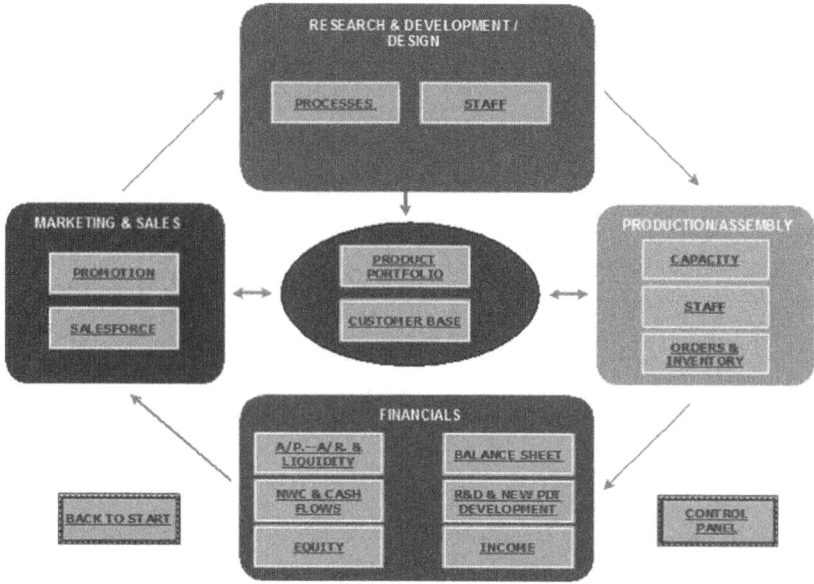

(d)

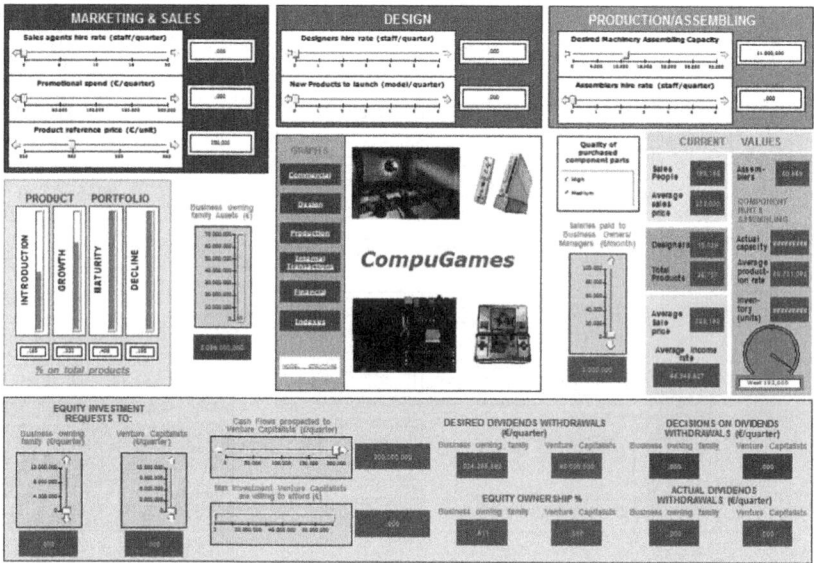

Fig. 5.5 (continued)

length of product life-cycle stages, product price, and component parts unit purchase cost.[4]

By changing input parameters, and replicating the same decisions previously
done with standard input settings, teams may perceive how the initial system
conditions may strongly affect performance.

The 'transparent box' stock and flow window allows participants to develop a
capability to perform a structure and behavior analysis, through simulation.

Figure 5.6a, b illustrate two examples of such model windows, related to product
portfolio and customer base dynamics, respectively.

Visualizing the stock and flow simulation model helps players to map the
feedback loops that are responsible for the experienced performance in the simulation runs.

For instance, Fig. 5.7 frames main feedback loops associated with R&D and
commercial investment policies aimed to foster business growth. The reinforcing
loop 'R1' shows how launching new products may increase the product portfolio
size. After a delay, related to the time it takes for new products to enter in a growth
stage, this will increase sales orders and—other conditions being equal—revenues.
This will contribute to increase cash flows and liquidity, which would allow the
business to fund new product launching.

The described loop 'R1', which triggers cash flows from new product development, can also be combined with two more reinforcing loops ('R2' and 'R3'),
which may further boost business growth. Such loops are, respectively, related to
the reinvestments of cash flows in improving sales force size and market support,
through promotional efforts.

However, the dominance of the described reinforcing loops fostering growth can
be counteracted by the balancing loop 'B': a too intensive new product launch rate
may significantly reduce the average product portfolio age. A too low product
portfolio age perceived by the market increases the promotional effort needed to
sustain new products. In this case, if the firm does not properly match its commercial and R&D budgeting policies, an R&D overinvestment and a commercial
underinvestment may generate a promotion gap (i.e. a difference between desired
and actual promotional efforts), which would reduce the sales orders, revenues, and
cash flows, in respect to the budget. Negative cash flows would reduce bank balances. This would tackle aggressive R&D policies, based on intensive new product
launching.

The 'control panel' window allows players to make their own decisions, each
quarter. It also supports them to quickly visualize the current value of a number of
critical variables (e.g.: those related to capacity and income), and to access time
graphs, related to different sectors. The 'time graph' windows portray strategic

---

[4]Teams can change such inputs only based on learning facilitators' instructions. If this occurs, the
same parameters must be changed in same way by all the teams. This is to enable, in a plenary
debriefing session, a comparative analysis across the simulation results obtained by the different
groups.

(a)

(b)

Fig. 5.6 **a** Product portfolio stock-and-flow structure. **b** Customer base stock-and-flow structure

resources, performance drivers and end-results that are periodically updated by the simulator after each quarter decision run.

After each simulation run, teams are asked by learning facilitators to draw a DPM chart by linking the time graphs one another, so to visualize and better frame the dynamics underlying business performance, and their causes.

At the end of each simulation run, the SD model automatically transfers results to a spreadsheet accounting model, which portrays the company financial statements (Fig. 5.8a, b, c). This allows the players to reframe the simulated system's behavior through an accounting view, which complements the DPM perspective and directly supports them to sketch a dynamic business plan.

Figure 5.9 shows some examples of pop-up windows that players may visualize while using the simulator. The purpose of such windows is to generate in a virtual

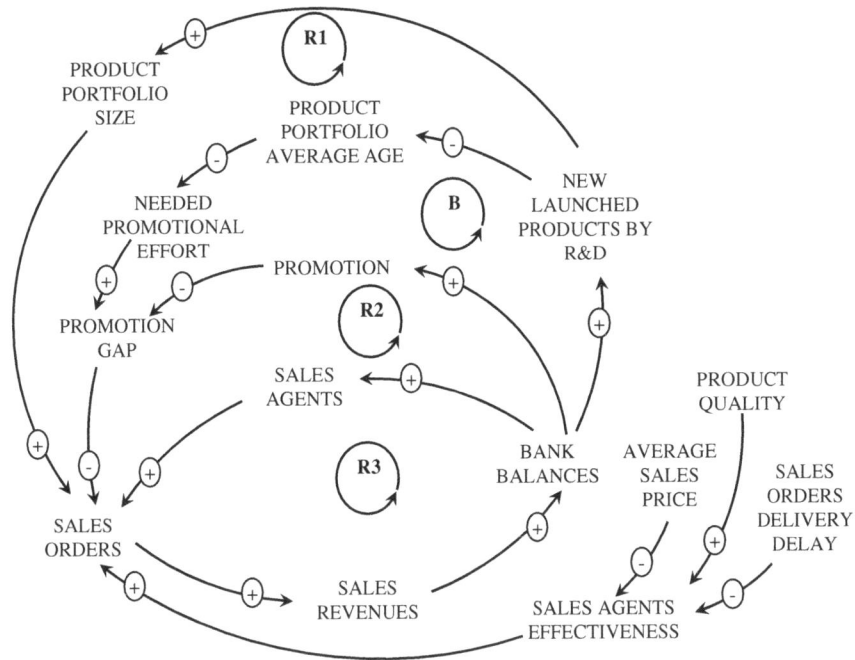

Fig. 5.7 Feedback loops associated with R&D and commercial investment policies

environment—such as the CompuGames *microworld*—an emotional pressure on decision makers about the perceptions of different stakeholders on business performance.

As Fig. 5.9 illustrates, warning messages may refer to issues like: lack of product promotional support, financial or capital shortage, overhead and structural costs dynamics, stakeholders' satisfaction. Also pop-up windows are used to stop simulation, if the company bankrupts due to bank debts higher than the maximum bank credit or to an equity deficit.

Teams should be able to use warning messages as a means to improve their own abilities to explain the causal structure underlying business performance and, perhaps, to adjust policies during a simulation run. However, such warning messages may also generate reactive and emotional decision-making, leading to worse performance. Analyzing the perils from emotional behavior and decision-making (and the related causes) is an important discussion topic in de-briefing sessions.

During the simulation, players can experience the level of complexity for *CompuGames* to survive as a family-owned, small business in its competitive and stakeholder systems. They can also perceive the long time it takes for the business to build up a product portfolio and a market image that may provide the basis for a future stable growth and profitability, and a strong liquidity position. They may experience that—in order to pursue business survival and development—most of the 4-years simulation time should be characterized by significant investments in

(a)

BALANCE SHEET (Euro)	Initial value	FIRST YEAR				%	SECOND YEAR				%
		1st Quarter	2nd Quarter	3rd Quarter	4th Quarter		1st Quarter	2nd Quarter	3rd Quarter	4th Quarter	
NET INVESTMENTS	18.957.849	13.240.741	10.106.567	8.618.957	8.417.511	100%	38.452.360	39.962.765	42.218.275	43.744.641	100%
Fixed assets:	8.111.667	8.170.107	8.213.340	8.253.680	8.291.094	98%	28.185.015	28.514.020	28.816.814	29.099.000	67%
- Assembling machinery	8.100.000	8.100.000	8.100.000	8.100.000	8.100.000	1	27.000.000	27.000.000	27.000.000	27.000.000	1
- Capitalised design costs	11.667	70.107	113.340	153.680	191.094	0	1.185.015	1.514.020	1.816.814	2.099.000	0
Working capital:	646.352	603.481	545.652	365.277	126.416	2%	10.267.345	11.052.056	9.780.323	9.842.086	22%
- End products inventory	112.500	54.286	50.995	47.903	24.549	0	1.618.285	1.886.640	1.632.505	1.740.952	0
- Component parts inventory	128.250	344.625	310.263	141.460	7.960	0	3.261.162	2.779.080	1.830.323	1.981.242	0
- Accounts receivable	405.602	204.570	184.395	175.914	93.907	0	5.387.899	6.386.336	6.317.495	6.119.892	0
Positive bank balances	10.199.831	4.467.153	1.347.575	0	0	0	0	396.689	3.621.138	4.803.555	0
EQUITY & LIABILITIES	18.957.849	13.240.741	10.106.567	8.618.957	8.417.511	100%	38.452.360	39.962.765	42.218.275	43.744.641	100%
Long term funds:	14.919.648	11.494.966	9.323.421	8.026.594	7.880.111	1	32.494.374	32.630.071	33.888.101	34.697.527	1
- Equity	6.819.648	4.292.800	2.820.051	1.787.391	668.587	0	14.388.890	16.129.126	18.052.428	19.379.643	0
invested by the family	6.819.648	4.292.800	2.820.051	1.787.391	668.587	8%	8.786.408	9.849.057	11.023.498	11.833.946	27%
invested by venture capitalists	0	0	0	0	0	0	5.602.487	6.280.069	7.028.930	7.545.697	0
- Long term bank debts	0	0	0	280.054	1.675.345	0	749.778	0	0	0	0
- Debts towards assembly machinery suppliers	8.100.000	7.202.166	6.503.370	5.959.489	5.536.179	1	17.355.706	16.500.945	15.835.673	15.317.884	0
Short term debts:	4.038.201	1.745.774	783.146	592.023	537.399	0	6.076.735	7.482.694	8.471.840	9.047.114	0
- Accounts payable	305.513	536.081	100.954	23.335	4.311	0	2.653.510	2.327.873	2.129.818	2.322.839	0
- for salaries	42.968	32.773	28.494	26.766	22.092	0	735.059	852.158	886.758	925.545	0
- for other industrial costs	305.220	317.048	331.187	330.247	323.987	0	642.864	702.366	761.012	859.596	0
- for promotional activities	3.200.000	675.342	138.073	28.229	5.771	0	155.666	172.871	167.711	34.288	0
- for administrative costs	184.500	184.530	184.438	183.446	181.238	0	1.587.241	2.098.573	2.516.614	2.810.729	0
- for taxes	0	0	0	0	0	0	302.397	1.328.852	2.009.928	2.094.116	0

(b)

INCOME STATEMENT (Euro)	FIRST YEAR					SECOND YEAR				
	1st Quarter	2nd Quarter	3rd Quarter	4th Quarter	TOTAL	1st Quarter	2nd Quarter	3rd Quarter	4th Quarter	TOTAL
SALES REVENUES	667.965	579.254	554.582	418.782	2.220.585	16.503.179	19.661.882	20.695.576	18.612.858	75.483.793
(-) COMPONENT PARTS PURCHASING COSTS:	1.704.741	709.780	259.899	171.540	2.845.956	9.003.978	9.028.867	8.800.850	7.969.854	34.553.549
Purchasing costs	1.923.117	675.418	91.092	56.040	2.725.666	9.947.597	8.546.785	7.952.092	8.150.773	34.497.248
(+) Change in component parts inventory	218.375	-34.363	-168.305	-135.500	-120.290	943.610	-482.081	-848.758	150.919	-356.301
FIRST CONTRIBUTION MARGIN	-1.036.776	-130.526	294.887	247.242	-625.573	7.499.201	10.833.515	11.894.728	10.632.802	40.650.244
(-) DIRECT INDUSTRIAL COSTS:	1.202.423	1.212.981	1.199.743	1.189.381	4.804.514	4.249.820	4.342.729	4.352.580	4.419.901	17.364.600
Handling	73.670	79.524	74.243	72.187	297.833	147.583	213.178	267.106	308.113	935.985
Setup	99.424	107.172	96.242	90.716	389.353	298.954	185.370	159.118	200.656	782.058
Assembling staff salaries	17.561	16.513	15.514	12.718	62.528	477.060	358.230	337.751	525.286	2.093.328
Assembling machinery depreciation	1.012.500	1.012.500	1.012.500	1.012.500	4.050.000	3.375.000	3.375.000	3.375.000	3.375.000	13.500.000
Other industrial costs	1.250	1.250	1.250	1.250	5.000	11.017	12.960	13.375	12.688	50.239
INDUSTRIAL CONTRIBUTION MARGIN	-2.239.201	-1.343.488	-905.061	-942.159	-5.428.887	3.249.381	6.390.786	7.542.376	6.202.901	23.285.444
(-) DIRECT COMMERCIAL COSTS:	190.378	32.058	30.428	45.948	298.833	1.334.376	1.615.901	2.503.176	2.056.844	7.399.798
Salaries and sales commissions	32.164	28.767	27.537	22.814	110.882	1.423.780	1.715.107	1.930.708	2.023.624	7.093.219
Promotion costs	100.000	0	0	0	100.000	197.917	168.750	208.333	141.667	716.667
(-) Change in end products inventory	-58.214	-3.291	-3.091	-13.334	-87.951	187.421	268.359	-254.133	108.447	410.089
INDUSTRIAL AND COMMERCIAL CONTRIBUTION MARGIN	-2.429.579	-1.579.545	-935.489	-988.107	-5.718.720	1.915.509	4.875.285	5.149.200	4.146.058	15.885.846
(-) DIRECT ADMINISTRATIVE COSTS:	81.643	81.580	80.629	73.980	322.833	1.294.876	1.476.522	1.656.681	1.717.314	6.067.373
Revenue cycle management costs	62.500	62.500	62.500	62.500	250.000	182.100	206.330	211.200	215.598	815.428
Expenditure cycle management costs	18.003	17.893	16.742	14.697	67.335	99.889	252.869	382.815	455.041	1.170.612
Depreciation on capitalised costs for new product design	1.141	1.187	1.567	1.783	5.698	9.513	17.848	30.485	48.799	106.649
Other variable administrative costs	0	0	0	0	0	945.374	999.275	1.032.162	997.876	3.974.686
NET INCOME BEFORE OVERHEAD COSTS	-2.513.223	-1.457.125	-1.016.118	-1.047.087	-6.051.553	678.429	3.198.765	3.512.537	1.428.744	9.818.475
(-) OVERHEAD COSTS:	13.625	13.625	16.541	51.717	99.508	363.278	1.458.330	1.580.235	1.301.529	4.512.372
Fixed administrative costs	13.625	13.625	15.625	15.625	62.500	15.625	15.625	15.625	15.625	62.500
Financial costs	0	0	916	36.091	37.008	7.547	18.078	0	0	26.624
Taxes	0	0	0	0	0	340.106	1.423.327	1.573.610	1.085.904	4.425.447
NET INCOME	-2.526.848	-1.472.750	-1.032.659	-1.118.804	-6.151.061	315.151	1.740.253	1.923.302	1.327.216	5.305.901

(c)

CHANGE IN COMMERCIAL NET WORKING CAPITAL - N.W.C. (Euro)	FIRST YEAR					SECOND YEAR				
	1st Quarter	2nd Quarter	3rd Quarter	4th Quarter	TOTAL	1st Quarter	2nd Quarter	3rd Quarter	4th Quarter	TOTAL
NET INCOME	-2.526.848	-1.472.750	-1.032.659	-1.118.804	-6.151.061	315.151	1.740.253	1.923.302	1.327.216	5.305.901
(+) DEPRECIATION	1.013.641	1.013.687	1.013.887	1.014.283	4.055.499	3.384.513	3.392.848	3.405.485	3.423.799	13.606.649
(=) INTERNAL FLOW OF FUNDS	-1.513.207	-459.063	-18.772	-104.521	-2.095.563	3.699.664	5.133.102	5.328.790	4.751.015	18.912.590
(+) Working capital increase:	1.246.438	1.448.663	1.234.761	889.464	1.419.323	40.309.350	49.728.422	50.931.168	47.353.000	191.786.200
- component parts costs	1.176.473	369.409	690.176	470.602	3.168.745	27.090.301	30.064.040	30.155.370	28.900.346	116.002.307
- accounts receivable increase	697.963	579.234	554.582	418.782	2.220.585	16.503.179	19.661.382	20.695.576	18.612.856	75.483.793
(-) Short term debts decrease:	4.695.341	2.007.711	693.394	624.178	7.706.624	13.578.329	12.871.369	12.607.181	12.743.336	31.800.393
- component parts suppliers payments	1.660.569	1.110.344	168.711	37.069	5.026.868	10.261.523	8.672.422	8.090.147	7.997.793	35.141.646
- salaries payments	119.521	93.981	98.506	79.303	379.051	1.219.020	2.301.081	2.797.141	2.841.108	10.226.250
- administrative costs payments	94.087	96.110	93.839	85.030	383.097	677.817	982.968	1.203.759	1.390.063	4.234.367
- promotion costs payments	2.824.859	367.369	108.364	23.458	3.384.229	204.384	182.794	205.180	135.420	725.771
- other industrial costs payments	345.516	188.806	170.674	170.421	675.419	317.170	190.308	380.934	421.018	1.466.280
(=) N.W.C. FINANCIAL NEEDS	6.557.779	3.456.375	1.868.154	1.513.642	13.175.850	37.122.358	62.596.681	63.559.307	60.296.318	245.366.293
(+) Working capital decrease:	1.589.309	1.306.481	1.413.138	1.123.324	5.509.261	41.295.081	48.943.751	52.233.079	47.481.239	189.918.109
- assembled component parts	1.020.512	907.061	830.070	627.336	3.406.981	29.618.311	30.179.766	31.459.963	28.470.990	116.225.719
- accounts receivable payments	968.397	399.420	563.063	500.789	2.592.177	15.441.770	18.653.945	20.764.416	18.920.299	72.690.990
(-) Short term debts increase:	2.998.814	1.045.083	442.271	369.334	4.255.821	14.006.371	13.219.512	12.933.565	13.376.020	33.559.248
- component parts costs	1.921.117	675.418	91.082	38.040	2.725.666	8.947.597	8.546.785	7.832.082	5.130.773	34.497.248
- salaries and sales commissions costs	109.226	89.702	84.377	74.500	359.183	1.230.350	2.420.180	2.501.741	2.879.893	10.332.116
- administrative costs	94.120	96.018	94.860	92.822	378.803	1.341.988	1.474.298	1.621.800	1.684.159	6.023.228
- promotion costs	100.000	0	0	0	100.000	197.917	168.750	208.333	141.667	716.667
- other industrial costs	172.344	183.948	171.734	164.163	692.186	397.560	409.309	429.399	513.420	1.766.282
(=) N.W.C. FINANCIAL COVERAGES	4.288.123	2.351.570	1.857.407	1.497.670	10.165.081	55.297.452	62.183.134	63.166.644	60.857.320	243.474.658
(a = s - i) N.W.C. FLOW	2.269.556	904.800	10.748	-16.237	2.980.366	1.804.907	435.457	-1.393.137	-570.929	113.238
(s = a - c) CURRENT CASH FLOW	-5.762.763	-1.363.383	-28.520	78.718	-5.076.429	5.504.757	4.696.824	6.916.927	3.322.004	18.890.313
(-) Change in fixed assets:	39.582	44.420	41.727	39.137	164.926	348.469	346.039	339.101	330.983	1.360.590
Fixed assets increase:	1.072.082	1.056.920	1.054.227	1.051.637	4.234.926	3.723.469	3.721.039	3.708.102	3.705.983	14.860.590
- assembly machinery acquisitions	1.012.500	1.012.500	1.012.500	1.012.500	4.050.000	3.375.000	3.375.000	3.375.000	3.375.000	13.500.000
- capitalised costs for new product design	59.582	44.420	41.727	39.137	184.926	348.469	346.039	333.101	330.983	1.360.590
(-) Long term debts increase (assembly machinery suppliers)	-1.012.500	-1.012.500	-1.012.500	-1.012.500	-4.050.000	-3.375.000	-3.375.000	-3.375.000	-3.375.000	-13.500.000
(+) Change in long term debts and equity:	-6.910.154	-1.714.296	-1.596.301	-1.433.010	-8.613.321	-4.170.766	-3.100.304	-3.399.186	-9.606.601	-14.541.368
Equity increase	0	0	0	0	0	81	3	0	0	83
(+) taxes debts increase	0	0	0	0	0	302.397	1.026.455	681.076	84.188	2.094.116
(-) long term debts payments	1.910.154	1.751.296	1.596.581	1.433.010	6.613.321	4.473.214	4.128.762	4.040.272	3.892.786	16.826.067
(-) Dividends payments	0	0	0	0	0	0	0	0	0	0
(=) NET CASH FLOW	-5.732.670	-9.118.379	-1.627.429	-1.283.291	-13.875.175	-2.630.478	1.546.467	3.224.448	1.182.418	2.897.355

Fig. 5.8 **a** Balance sheet simulation results. **b** Income statement simulation results. **c** Flow of funds simulation results

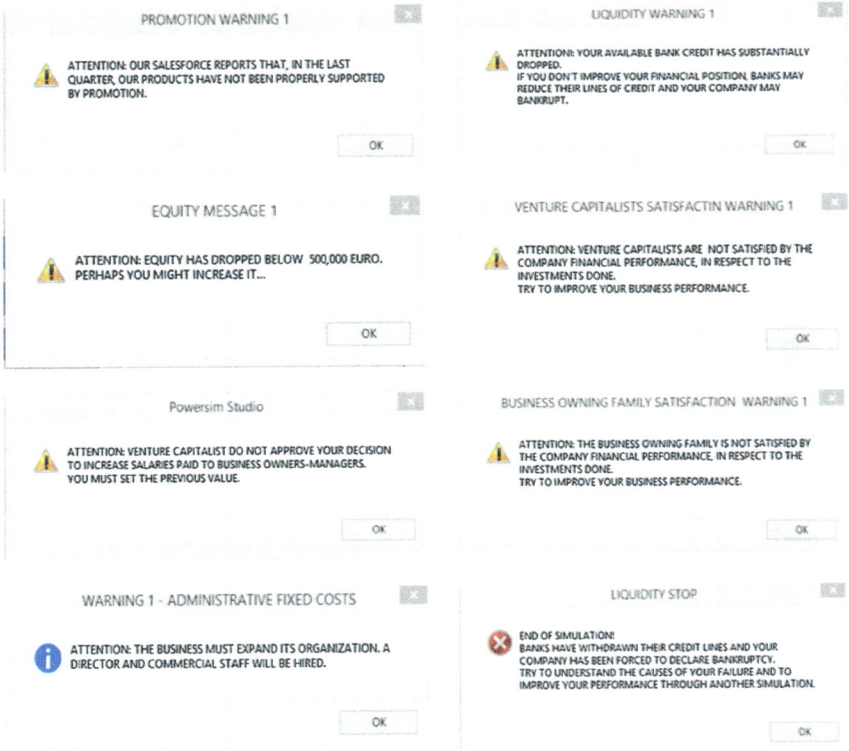

Fig. 5.9 CompuGames simulator's pop-up windows

R&D, production/assembling, and commercial subsystems, to build a set of strategic resources generated by company routines, whose deployment may ensure future profitability, liquidity and a stable market share.

The benefits of such learning process based on "virtual experience" can be substantial to foster a major change in business-owners/entrepreneurs' mental models. Company failures often result from an inclination of their owners-founders to underestimate the payback time for their investments and the financial needs associated with the development of those strategic resources that will support sustainable growth. Regarding this issue, business owners-founders are often inclined to underestimate the risks on the future business liquidity and profitability, related to an excess in the resort to bank credit. By playing the CompuGames simulator, participants may experience the effects of the reinforcing loops produced by a too high "debts-to-equity" ratio, and the related risk of company bankruptcy.

In the CompuGames case-study, contracting such risks may require the need to accept that both the business-owning families and external funders (such as venture capitalists) will invest into the business, perhaps through a substantial equity increase. This policy may not be 'pain-free' for the business owners-entrepreneurs. In fact, it may imply for them a much lower authonomy in decision-making on

(a)

(b)

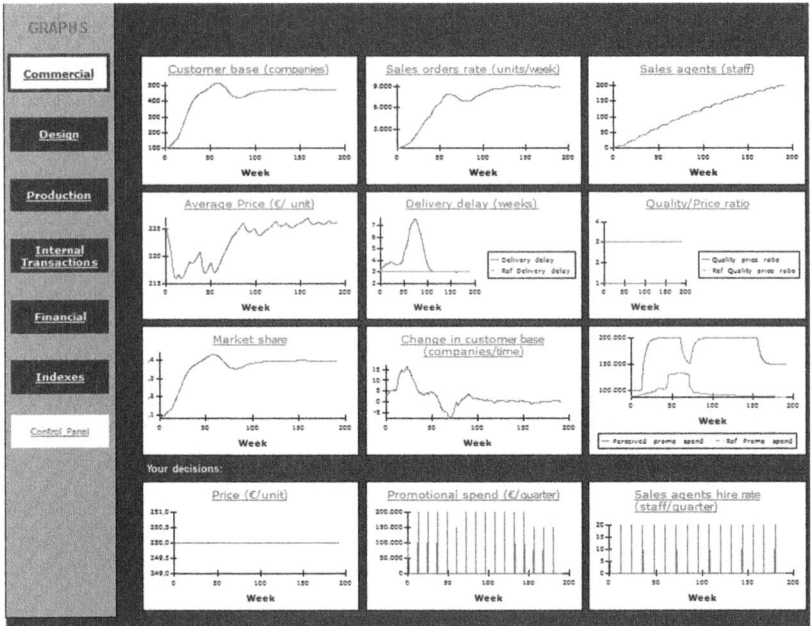

Fig. 5.10 **a** An example of decision data from a simulation. **b** An example of decision data from a simulation (commercial graphs). **c** An example of decision data from a simulation (R&D/product design graphs). **d** An example of decision data from a simulation (production/assembling graphs). **e** An example of decision data from a simulation (financial graphs). **f** An example of decision data from a simulation (performance index graphs)

business governance and strategic management. They might need to learn and practice new skills to communicate and share with venture capitalists their own business ideas. They may also need to adopt a professional approach in recruiting human resources: for instance, involving family members in business management and governance would not be an automatic or easy option.

Figure 5.10 provides an example of a simulation run leading to a sustainable business growth rate, which is compatible with profitability and liquidity. This policy aims to substantially increase the company size. Growth is pulled by strong investments in promotional spend and constant sales agents and assembly staff hire rates. Product innovation and launching is key in the adopted policy.

(c)

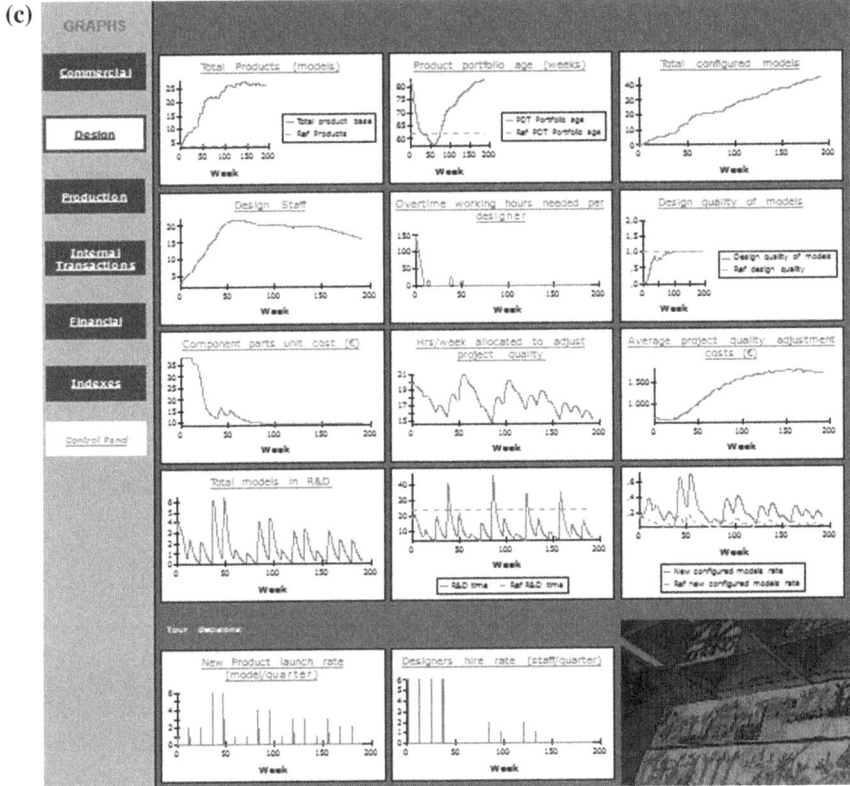

(d)

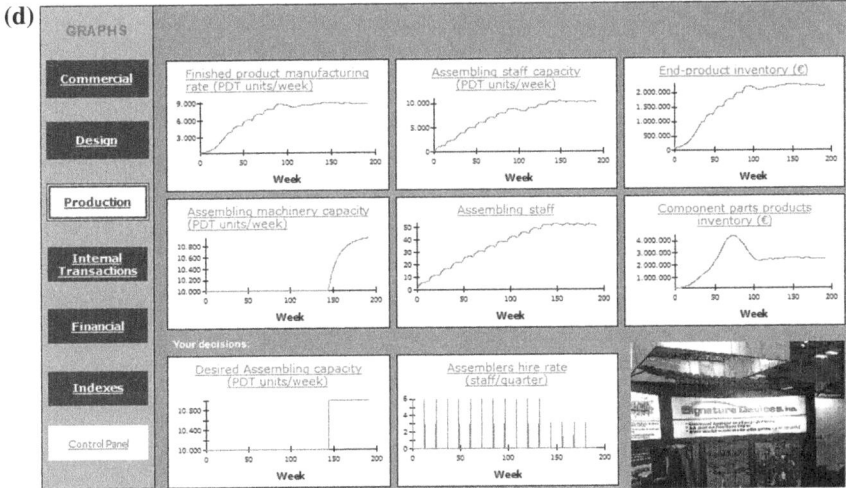

Fig. 5.10 (continued)

(e)

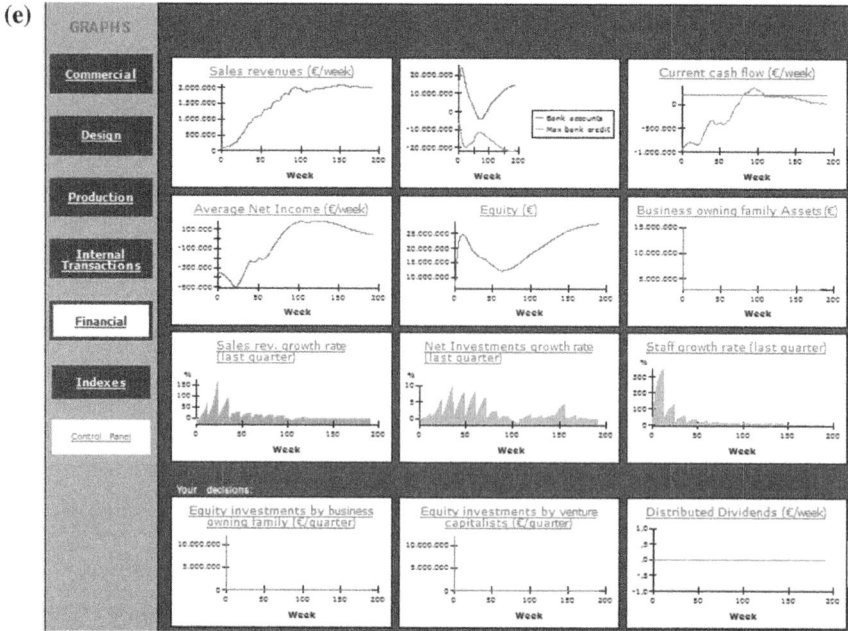

(f)

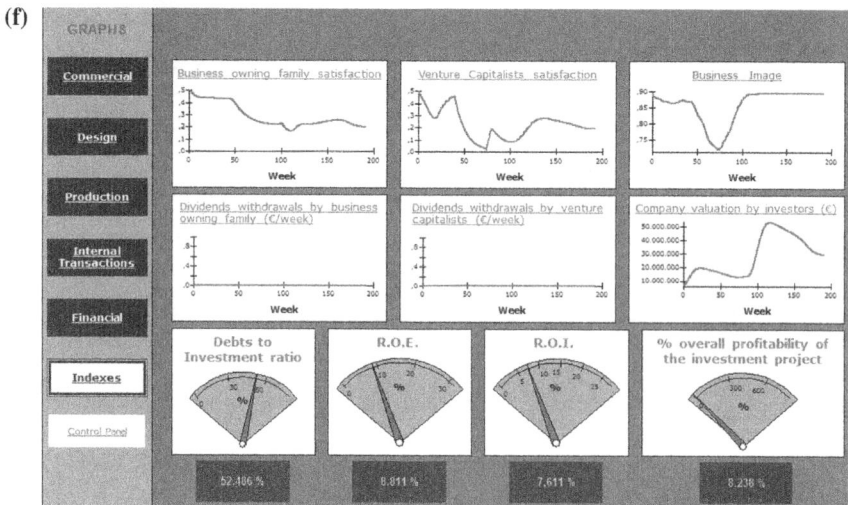

Fig. 5.10 (continued)

The graphs related to the commercial, product design, production, and financial subsystems show that it takes about two years for the company to turn into a profitable and cash flow generating business. This requires that a 'no-dividends' policy is adopted, to sustain the high growth rate of the firm: in the 4-years the business experiences a 600 % increase in its "product portfolio", a 500 % increase in its customer base, and a 400 % increase in market share. Though such policy may generate good results in terms of growth and an acceptable profitability and liquidity in the long run, and the business capital value increases by about 300 % in the four simulation years, it reduces the level of financial stakeholders' satisfaction.

5.3 Modeling Commercial and Financial Subsystems Through DPM: *Licari & Sons* Case-Study

A second example of a SD model framing the interplay of financial and commercial processes in a small family firm will be illustrated in this section using the DPM perspective.

The model was applied to provide an initial basis for analysis and discussion with the two owners and the chartered accountant of *Licari & Sons*, a small company operating as a regional wholesaler in the pharmaceutical industry. The two brothers owning the business were simultaneously involved in strategic and current decision-making. One of them was responsible for commercial activities, the other managed relationships with banks. The market was characterized by the strong bargaining power of large producers and a fragmented, standardized wholesale supply, involving a strong competition on price, terms of payment allowed to customers and delivery time.

The industry had been recently affected by unexpected structural changes, related to the reduction of funds granted by the State to pharmaceutical companies and the drastic reduction of State financial contributions to citizens for the purchase of medicines.

In a few years, such events led to a decrease in demand and were a primary cause of crisis in many firms operating both in the production and distribution stages. In the regional market where *Licari & Sons* operated, an increasing number of pharmacies started to shift their financial difficulties on wholesalers. They did this by postponing payments of purchased goods beyond negotiated terms. On the other hand, the fear of market share loss exacerbated competition between wholesalers; at the same time terms of payment negotiated with producers were decreasing.

Like its competitors, *Licari & Sons* started to pursue a commercial strategy aimed to keep (and, possibly, to increase) its customer base and market share. This was accomplished through a progressive rise in price discounts and terms of payment on sales. Very often, when a new order was submitted, terms of payment were re-negotiated with clients by phone. One of the arguments used by clients to persuade the firm to grant longer terms of payment was the availability of other wholesalers to allow them higher payment delays. Clients also made emotional pleas pressuring Mr. Licari, asking him to postpone the collection of their accounts payable, in the name of their old commercial relationships.

The described policy gave rise to a sharp increase in both sales revenues and income. It also generated a sharp financial crisis, which was detected by the firm

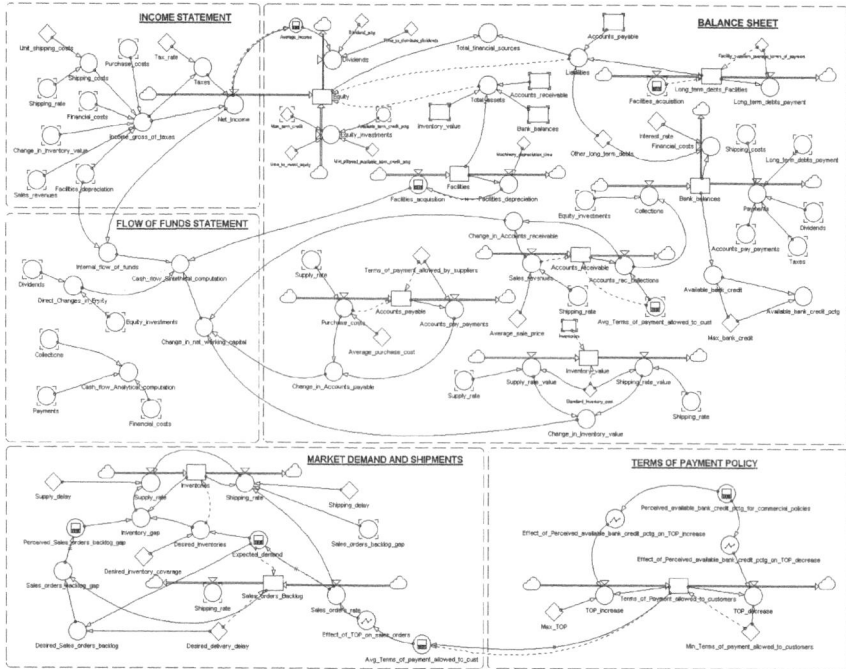

Fig. 5.11 A SD model framing financial and commercial subsystems

only after banks started to increase their pressure on Mr. Licari. This was done by asking him to reduce the firm's negative balances and to submit a business plan to prove company solvency.

No formal planning was done by the firm. The only tools used were transaction systems (e.g., inventorying, invoicing, etc.) and financial accounting.

In order to help the business owners to frame dynamic relationships, the firm's chartered accountant supported a group of modelers to calibrate a simple SD model (see Fig. 5.11) to past financial statements.

The model depicts cash flow as the net change in bank balances. Current cash flows result from the difference between internal flow of funds (i.e., income gross of depreciation) and the change in commercial net working capital (i.e., inventories + accounts receivable − accounts payable). A higher income increases current cash flows if the change in net working capital (corresponding to an extra current financial need) is lower than the increase in internal flow of funds. Furthermore, current cash flow becomes negative if the change in net working capital (NWC) is higher than the internal flow of funds. The total cash flow portrayed in the model is calculated by adding to the current cash flow the direct change in equity (i.e., investments minus profit withdrawals) and deducting monetary needs associated with payments for machinery replacement. Total cash flow can be also analytically calculated, i.e. from the algebraic sum of different flows impacting on bank accounts (Accounts receivable collections—Financial costs—Payments referred to accounts payable on purchased goods, shipping costs, long-term debts for machinery acquisition and dividends).

Relationships were framed between terms of payment policies, income, commercial NWC, and cash flows in order to understand limits to growth associated with the company financial structure and demand elasticity. The model calibration process gradually involved the two business owners, as they started to be questioned on issues like: retailers and competitors' reactions to changes in terms of payment on goods sold, perception delays in available bank credit, etc.

Figure 5.12 frames the SD model through a DPM view. The figure shows that a change in the terms of payment allowed to customers affects two main performance drivers: (1) the terms of payment ratio, and (2) the net change in accounts receivable and accounts payable.

The first performance driver is the ratio between the company's average terms of payment and competitors' terms of payment allowed to customers. A ratio of '1' would imply no effect of terms of payment policies on sales orders. A ratio higher than '1' implies higher sales orders for the company (in respect to the standard rates); a ratio lower than '1' implies lower sales orders rate than the standard. The effect of such performance driver on sales order (end-result) is modeled through a sales orders multiplier.

The second performance driver is the ratio between the actual and budgeted difference in accounts receivable and payable from a period to another. An increase in such difference implies an increase in financial needs, which involves—other conditions being equal—an increase in the NWC and a decrease in the current cash flows (end results).

Figure 5.12 also shows that sales orders are co-flows of other lower level end-results, i.e.: sales revenues, which in turn affect the current income, which affects the current cash flow.

Cash flow accumulates into the stock of bank accounts (i.e. liquidity); income accumulates into the equity stock. Such stocks are strategic resources which feedback into business performance through another driver, i.e.: the "debts-to-equity" ratio. In fact, a negative cash flow will reduce bank balances. If bank balances are negative, a reduction in their level will imply an increase of current debts by the firm. Likewise, a negative income rate (i.e. a loss) will reduce equity. Under both viewpoints, a negative financial performance will increase the "debts-to-equity" ratio (in respect to the target or budget standard). This will generate higher financial costs, due to both the higher debts and to the higher interest rate that a lower company solvency perceived by banks may imply.

The described DPM view can be re-framed through the feedback loop diagram illustrated in Fig. 5.13.

The figure shows the effects of gradual—but continuous—increases in the terms of payment allowed by the business to its clients, in order to increase sales revenues. As customers are sensitive to payment delays and the unit contribution margin on goods sold is positive, such policy can successfully increase the income rate. A higher income also raises bank balances, provided that cash flows are

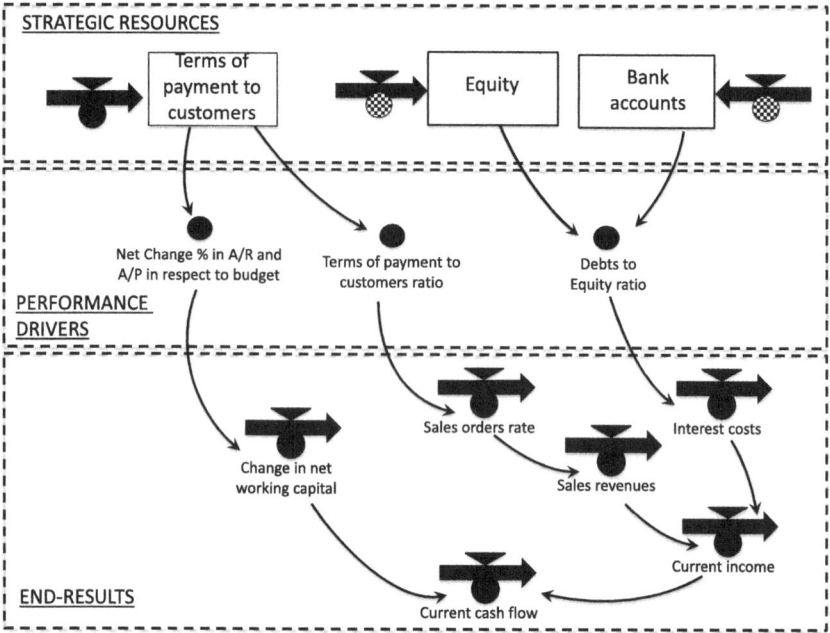

Fig. 5.12 A DPM view of the effects of terms of payment policies on sales orders, liquidity and income

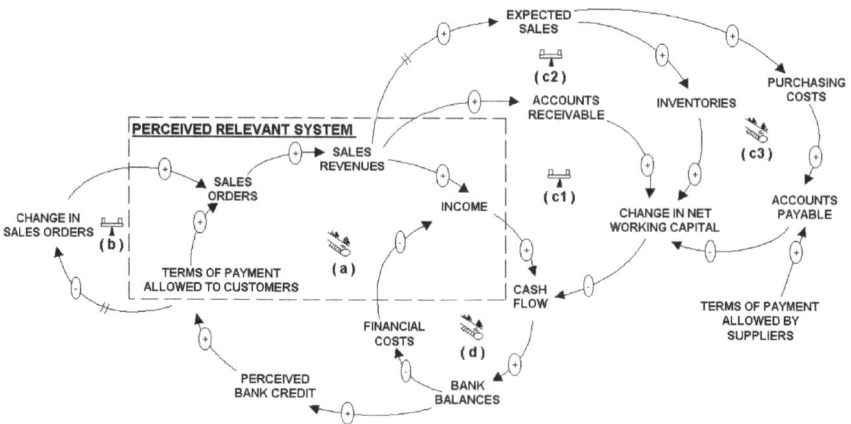

Fig. 5.13 Feedback structure underlying income, commercial net working capital and current cash flow behavior associated to terms of payment policies

positive. The increase in bank balances raises perceived bank credit, thereby encouraging the firm to gradually boost again terms of payment allowed to customers (reinforcing loop "a").

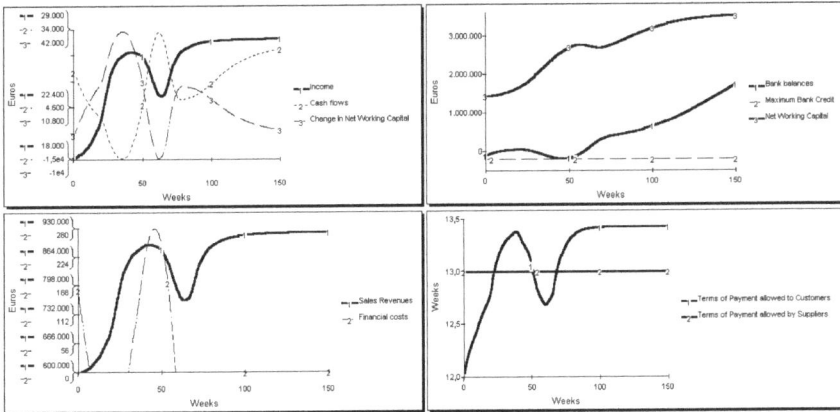

Fig. 5.14 Behavior of key variables generated by changes in terms of payment allowed to customers

There are, however, two main limits to growth to the sales revenues and income rates, i.e.:

(1) Demand elasticity to terms of payment;
(2) Changes in the company financial structure caused by higher sales revenues.

Concerning the first limit, the model implies that effects generated on sales orders by further increases in terms of payment allowed to customers can be depicted by an 's-shaped' curve. As shown in Fig. 5.14, this means that, when the commercial policy lever is increased beyond a threshold level (in this case, about 13 weeks) the change in sales orders will decrease (balancing loop "b").

Concerning the second limit, higher sales revenues will increase financial needs for both inventories and accounts receivable (the last effect is also amplified by higher terms of payment allowed to customers). This raises the NWC, thereby causing lower cash flows. If the change in the NWC is higher than income gross of depreciation, cash flows are negative. This leads to a reduction in bank balances, perceived bank credit and terms of payment allowed to customers (balancing loops "c1" and "c2"). Such a limit to growth could be counterbalanced by higher terms of payment allowed by suppliers on purchased goods, giving rise to higher accounts payable and—other conditions being equal—lower NWC, higher cash flows and bank balances (reinforcing loop "c3").[5]

---

[5]Another possible way to finance sales growth could be associated with new liquidity investments as equity, made by business owners.

The above balancing loops prevail over the reinforcing loop "a" after about the 20th week, when bank balances start to drop. However, such a limit to growth is not immediately perceived by the company, which continues to raise terms of payments allowed to customers until about the 40th week. This policy both determines a lower increase in sales revenues (caused by demand elasticity) as well as decreasing cash flows and bank balances (see Fig. 5.14). An unintended side-effect of this policy is also associated with higher financial costs on negative bank balances, leading to lower income and cash flows which reduce bank balances again (vicious reinforcing loop "d").

When the firm realizes the above limits to growth and risks they imply for profitability and solvency, it starts to reduce terms of payment allowed to customers. If this happens before business profitability has been prejudiced due to the effects produced by loop "d" and if market permits, such policy allows the business to attain a NWC reduction and an increase in cash flows and bank balances, although both sales revenues and income decrease.

After about the 60th week the above policy has allowed the firm to improve its financial structure, so that it can rely on a positive available bank credit, which fosters a new gradual increase in terms of payment allowed to customers, that makes loop "a" dominant again, until the balancing loop "b" stabilizes the system.

Although the above relationships might appear commonplace, if only observed through a *post facto* perspective, the dynamics they are likely to generate are often counterintuitive and puzzling for many small business entrepreneurs. Among main factors explaining such perception difficulties are:

- a counterintuitive behavior of key variables such as income, cash flows and change in the NWC;
- inertial effects generated by decision makers' policies, due to delays embodied in the relevant system;
- SME entrepreneurs' high emotional involvement in current operations, which makes difficult to perceive how continuous small changes in the short run are likely to generate structural modifications in the relevant system's structure;
- a static and discrete view of business phenomena often provided by accounting reports to SME entrepreneurs;
- a weak relative weight of the firm in its market, especially towards suppliers and distributors.

Figure 5.15 portrays accounting reports whose values are generated by the SD model simulations previously discussed.

Embodying accounting variables in a SD model allows one to open the entrepreneur's mind on the processes generating forecasted and actual values depicted in a balance financial statement. For instance, rather than focusing only on income, internal flow of funds, change in the NWC and cash flows, decision makers can be

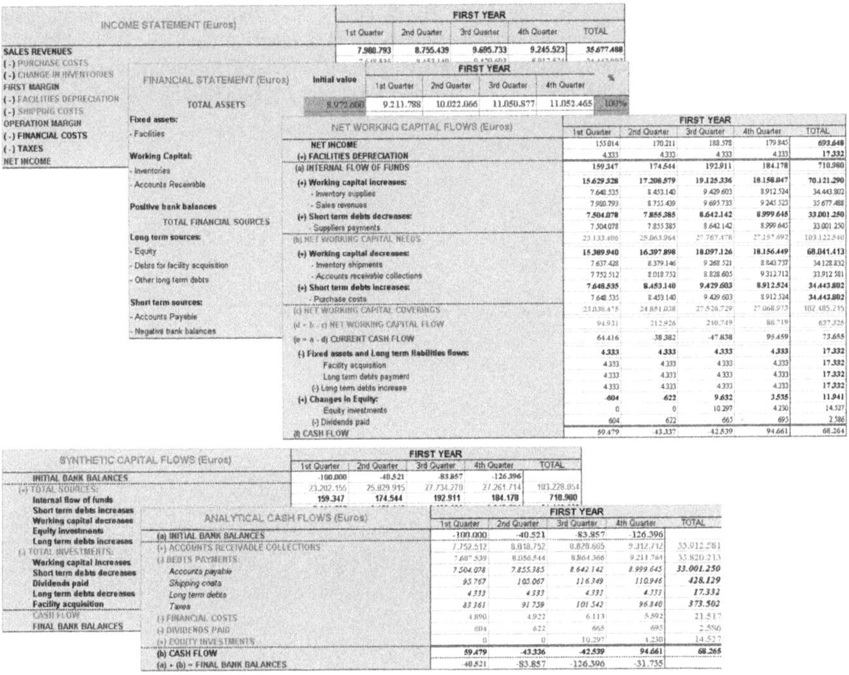

Fig. 5.15 Accounting reports generated by the SD model simulation results

supported by SD models to understand policy levers on which to act in order to affect key variables' performance over time, according to a desired direction.

References

Davidsen, P., Spector, J., & Morgan, K. (2000). System dynamics and interactive learning environments in simulation and gaming. *Special Issue: Simulation & Gaming, 31*, 2.

Ford, A. (1996). Testing the snake river explorer. *System Dynamics Review, 12*(4), 305–330.

Goosen, K., Jensen, R., & Wells, R. (2001). Purpose and learning benefits of simulations: A design and development perspective. *Simulation & Gaming, 32*(1), 21–39.

Grossler, A. (2004). A content and process view on bounded rationality in system dynamics. *Systems Research and Behavioral Science, 21*(4), 319–330.

Mohapatra, P., & Saha, B. (1996). System-dynamics-based game for new product growth. *Simulation & Gaming, 27*(2), 238–260.

Morecroft, J. (1987). System dynamics and microworlds for policymakers. *European Journal of Operational Research, 35*(3), 301–320.

Morecroft, J. (2007). *Strategic modeling and business dynamics*. Chichester: Wiley.

Morecroft, J., & Sterman, J. (eds.). (1994). *Modeling for learning organizations* (pp. XVII–XXII). Portland: Productivity Press.

Romme, A. (2004). Perceptions of the value of microworld simulation: Research note. *Simulation & Gaming, 35*(3), 427–436.

Rouwette, E., Fokkema, E., Van Kuppevelt, H., & Peters, V. (1998). Measuring marco polis management game's influence on market orientations. *Simulation & Gaming, 29*(4), 420–431.

Winch, G., & Arthur, D. (2002). User-parameterized generic models: A solution to the conundrum of modelling access for SMES? *System Dynamics Review, 18*(3), 339–358.

Index

A
Administrative product, 104, 119, 120, 138, 181
Auxiliary variables, 24, 28, 133

B
Balanced scorecard, 55, 66, 67, 144, 149, 154, 166, 169
Business management, 200, 222

C
Capacity resources (modeling), 107
Cash flow, 4, 35, 42, 55, 72, 74–76, 105, 110, 128, 130, 145, 146, 148, 149, 174, 194, 195, 208, 210, 212, 217, 224, 226, 227, 229, 230
Citizen satisfaction, 64, 154, 184
CompuGames simulator, 213, 214, 221
Conceptual stock-and flow models, 23
Controller (role of), 32, 33

D
Double-loop learning, 9, 10, 36, 38, 39, 41, 201
Dynamic balanced scorecard, 83, 167
Dynamic business plan simulator, 201, 203, 211, 213, 214
Dynamic complexity, 1–3, 14, 21, 32, 35, 37, 41, 44, 55, 56, 61, 65, 66, 71, 133, 199, 213, 214
Dynamic performance management
  and balanced scorecard in opera houses, 165
  and e-government, 184, 185
  and local area strategic planning, 13, 66
  and microworlds, 199–201
  and the mosaicoon case-study, 104, 105
  and the saturday evening post case-study, 94
  and user satisfaction in the public sector, 178
  chart, 26, 85, 94, 95, 100, 106, 146, 156, 160, 163, 169, 173, 218
  in crime reduction policies, 4, 22, 53
  in garbage collection, 4, 83, 85, 150, 151
  instrumental view, 72, 81, 86, 88, 99, 104, 105, 115, 117
  in the banking industry, 17, 56, 88, 176, 189, 195, 202
  in the telecom industry, 88
  in university institutions, 105, 119
  objective view, 117, 120, 121, 136
  subjective view, 117, 135, 137

E
End-results, 43, 66, 67, 72–76, 81, 83, 85, 89, 93–95, 99, 102, 105–107, 114–117, 124, 135, 137, 143, 145, 155, 177, 192, 195, 212, 218, 227
Entrepreneurial learning, 199, 201
External perspective in system dynamics modeling, 36, 41

F
Feedback and feed forward processes in planning & control, 12, 31, 159
Feedback loop, 3, 4, 11, 14, 17, 26–28, 39, 41, 42, 44, 68, 73, 85, 132, 192, 217, 227
Financial resources (modeling), 72, 107, 110, 117, 149, 172, 194, 208
Flow variables, 23, 24, 75, 81

© Springer International Publishing Switzerland 2016
C. Bianchi, *Dynamic Performance Management*, System Dynamics for Performance Management 1, DOI 10.1007/978-3-319-31845-5

9 783319 318448